Clinical Light Damage to the Eye

David Miller
Editor

Clinical Light Damage to the Eye

Foreword by Linus Pauling

With 85 Figures, 50 in Full Color

Springer-Verlag
New York Berlin Heidelberg
London Paris Tokyo

David Miller, MD
Associate Professor of Ophthalmology
Harvard Medical School
Ophthalmologist-in-Chief
Beth Israel Hospital
Boston, Massachusetts 02215
USA

Library of Congress Cataloging-in-Publication Data
Clinical light damage to the eye.
   Includes index.
   1. Eye—Effect of radiation on. 2. Light—
Toxicology. I. Miller, David, 1931–
[DNLM: 1. Eye Injuries. 2. Light—adverse
affects. WW 525 L723]
RE48.C641   1987   617.7′13   86–28063

Typeset by Arcata Graphics/Kingsport, Kingsport, Tennessee.
Printed and bound by Universitätsdruckerei Stürtz AG, Würzburg, Federal Republic of Germany.
Printed in the Federal Republic of Germany.

9 8 7 6 5 4 3 2 1

ISBN 0-387-96451-7 Springer-Verlag New York Berlin Heidelberg
ISBN 3-540-96451-7 Springer-Verlag Berlin Heidelberg New York

Foreword

To my mind, the superoxide radical discovered by Linus Pauling more than 50 years ago is about to become a major issue in American medicine. Uncannily, Pauling's early focus on vitamin C has pointed the way to the whole catalogue of free-radical scavengers, which we in medicine will be using in the coming decade. In ophthalmology, the basic scientists have been talking about the role of free-radical induction by light for some time. They have accumulated an increasing amount of evidence supporting the idea that prolonged light exposure contributes to cataract development and retinal degeneration. Through *Clinical Light Damage to the Eye*, we hope to bring this message to the practicing ophthalmologist.

Because Dr. Pauling's work bears so strongly on the key issue of free-radical damage, and because of my own great respect for him as a scientist and a man of rare courage, I invited Dr. Pauling to write the foreword to *Clinical Light Damage to the Eye*, which follows.

David Miller

* * *

The eye is a marvelous organ, permitting us to be aware of the nature of our environment. It might be considered to be the most marvelous of the organs. The heart, for example, is a rather simple pump, moving blood to the lungs to be oxygenated and then to the rest of the body to carry out its work of delivering oxygen and bringing waste products back to the places where they can be excreted. To a chemist, the liver might seem to be the most marvelous organ because of its thousands of different chemical factories, in which, with the aid of enzymes, many of the biochemical reactions required for life are carried out. But we see with the eye; it is obviously a marvelous organ.

The eye is also the most sensitive of the organs. Other organs can be severely damaged and still function well. The eye must retain its crystal clarity in order to be effective, and even slight damage can seriously impair vision.

Light itself, especially the ultraviolet light of that portion of the solar spectrum that reaches the surface of the earth, can do damage by its interaction with the many different kinds of molecules in the eye. People are blinded not only by accident but also by glaucoma, an increase in intraocular pressure, and by cataract formation. A great deal of human suffering could be avoided if scientists were to discover the mechanisms through which the eye is damaged by light, and then to find ways of preventing this damage.

This scientific problem is a difficult one. There have been some successes. In some countries the principal cause of blindness is a deficiency of vitamin A in the diet. It has been found that the addition of vitamin A to the diet not only prevents this blindness, but also improves the general health of the children to which it is administered. There is the possibility that nutritional factors, especially the vitamins, can be used to control glaucoma, cataract formation, and other diseases of the eye.

In recent years many able scientists have been carrying out careful studies of the eye, and there has been great progress in the understanding of the mechanisms by means of which light itself damages the eye. The results that have been obtained are discussed in this book, *Clinical Light Damage to the Eye*. The book should be of great help to investigators in this field and should lead to further progress. In addition, it contains much information of interest to readers other than specialists on the eye.

Linus Pauling
Palo Alto, California

Preface

The mammalian eye has been present and functioning on this planet for well over 125 million years. During this period, many adaptations that allow the animal to maximally meet its visual demands in its particular environmental setting have taken place. Modern man is a relative newcomer by the evolutionary timetable. Although a newcomer, man can be found in just about ever corner and crevice of the planet, subject to every conceivable variable of climate, temperature, diet, and sun exposure.

Man not only has had a relatively short time span for ocular adaptation, but has added stresses not commonly seen in the animal kingdom. For example, man has doubled his life span, thus forcing the original visual system to run efficiently for the equivalent of two original lifetimes. Modern man has continued to migrate from his ancestral settings. Thus, a light-skinned, blue-eyed individual, best adapted for cool northern climates, may retire to a sunny tropical setting, where light levels may be double that in the north. Modern man takes drugs that may make him more vulnerable to light toxicity. He also may be subject to dietary changes and deficiencies that may diminish his resistance to light damage. Happily, most of us are able to defend our eyes against light damage, but some of us cannot. Increasing evidence suggests that a number of eye diseases, including cataract and senile macular degeneration, may be related to light toxicity in the vulnerable patient.

In this book, we hope to give the clinician a clearer understanding of how light may harm the various portions of the eye, as well as suggestions for the protection of the vulnerable eye.

David Miller

Contents

Protecting the Eye from Light Damage

Overview of Light Damage to the Eye

Contributors

P. John Anderson, PhD, Howe Laboratory, Massachusetts Eye and Ear Infirmary; Assistant Professor of Ophthalmology, Harvard Medical School, Boston, Massachusetts, USA

Jeffrey D. Bernhard, MD, Associate Professor of Medicine; Director, Division of Dematology; Director, Phototherapy Center, University of Massachusetts Medical School, Worcester, Massachusetts, USA

David L. Epstein, MD, Director, Glaucoma Clinic, Massachusetts Eye and Ear Infirmary; Associate Professor of Ophthalmology, Harvard Medical School, Boston, Massachusetts, USA

Louis Erhardt, Illumination Engineer, Camarillo, California, USA

Sidney Lerman, MD, Professor, Department of Ophthalmology, Emory University School of Medicine, Atlanta, Georgia, USA

David Miller, MD, Associate Professor of Ophthalmology, Harvard Medical School; Ophthalmologist-in-Chief, Beth Israel Hospital, Boston, Massachusetts, USA

Linus Pauling, PhD, Research Professor and Chairman of the Board, Linus Pauling Institute of Science and Medicine, Palo Alto, California, USA

Robert Stegmann, MD, Professor and Chairman, Department of Ophthalmology, Medical University of Southern Africa, Pretoria, South Africa

John Weiter, MD, PhD, Director of Clinical Services, Eye Research Foundation; Assistant Professor of Ophthalmology, Harvard Medical School, Boston, Massachusetts, USA

Seymour Zigman, PhD, Professor, Department of Ophthalmology, University of Rochester School of Medicine, Rochester, New York, USA

*The Nature of Light and
of Light Damage
to Biological Tissues*

1

Radiation, Light, and Sight

Louis Erhardt and David Miller

Papers entitled "Retinal Light Exposure from Ophthalmoscopes, Slit Lamps, and Overhead Surgical Lamps," [1] "Ophthalmic Lasers, Photocoagulation, Photoradiation and Surgery," [2] as well as papers on the lighting conditions [3] for visual acuity testing and contrast sensitivity testing compose a significant portion of our ophthalmologic reading. These papers freely use unfamiliar terms such as watts, joules, irradiance, foot-candles, lux, lumens, candelas, millilamberts, etc. Of course, there are a number of excellent texts which rigorously define these terms. [4–6] Unhappily, if the terms and units are not used in almost a daily fashion, the meanings blur.

In this review the combined efforts of an illuminating engineer [*] and an ophthalmologist attempt to (a) logically group the units into categories that can be more easily remembered, (b) define the units of each category in user-friendly terms, and (c) give clinical examples of the use of the terms. It is hoped that this review will function as a useful reference guide.

Let's start by looking at the different ways that light can be analyzed and measured. Sensation is the mind's link with the physical world. Without this junction the world of matter and the world of the mind would remain forever separate. Matter and physical science exist as though man were left out. The world of matter and energy is colorless, toneless, neither warm nor cold; its spaces

---

[*] Louis Erhardt has designed the lighting for many Broadway shows, operas, theaters, the cabin of the Lockheed Electra, and Los Angeles International Airport.

are always of the same extent, time always of the same duration, mass invariable. Light in a physics textbook is electromagnetic radiation; sound, a vibratory motion in air or water; heat, a dance of molecules—all independent of man. Psychology is the science of the world as it exists in man's experience. It contains colorful pictures, harmonies, and melodies in sound, rough- and smooth-feeling surfaces. Space may be large or small, time ever so short or agonizingly long—there are no absolutes; all is variable. Psychology contains thoughts, emotions, memories, imaginations, and hopes, all of the considerations one ascribes to the mind.

Between the physical world and that of the mind lies an area where the two are purposefully related to each other—the psychophysical world. Here a *judgment* of the magnitude of sensation is coupled with a *measurement* of the physical cause of the sensation. This procedure leads to numbers, scales, and graphs, which form the basis of photometry. Since light is radiant energy capable of producing visual sensation, photometric measurements are made either by the human eye or are based on its visual response.

The power of the visible radiation obviously can be measured,[†] and thus we can express the power of the visible radiation of a mercury lamp or an incandescent lamp in watts. But the power of visible radiation is not proportional to the physiological effect and thus not a measure thereof. One watt of green radiation gives many times as great a physiological effect, i.e., more light, than does a watt of red or violet radiation. It also gives a different kind of physiological effect: a different color.[7]

Physical Radiation

To understand radiation units, it is best to measure them as strictly energy units with no link to physiology. The important units are: radiant energy, irradiance, radiant exposure, or dosage. A table of terms (Table 1.1) is presented which differentiates among the three disciplines, physics, psychophysics, and psychology. These correspond with the major sections of this chapter and it is recommended that reference to this table be made constantly as its terms are addressed.

[†] A black body of one square centimeter area operating at 2,045° Kelvin radiates 99.15 watts total; 1.66 watts between wavelengths 0.38 and 0.76 micrometers; and produces (by definition) 188.58 lumens.

Table 1.1. Terms and units of measurement.

Physics (radiation)	Psychophysics (light = photometry)	Psychology (visual perception)
Intensity of source, $W \cdot sr$	Intensity of light, cd/sr	Brightness of source
Irradiance on surface, W/cm^2	Illuminance on surface, lm/m^2	Brightness on surface
Radiance of surface, $W/sr \cdot cm^2$	Luminance of surface, $cd/sr \cdot cm^2$	Brightness of surface
Radiant emittance, W/cm^2	Luminous emittance, Lm/m^2	(Numbers, when they occur, are ordinal only)
Radiant energy, $J = W \cdot s$	Luminous energy, $lm \cdot hr$	
Radiant exposure, J/cm^2		

Key: W = watt, cm = centimeter, J =joule, sr = steradian, s = seconds, cd = candela, lm = lumen, hr = hour, m = meter.

Radiant Energy

Planck demonstrated that light energy varies inversely as the wavelength of light. The following units are used to describe wavelengths of visible light: Angstroms (A), nanometers (nm), microns, or micrometers (μm); 1A = 10nm = 10,000 μm. Visible spectrum: 4,000 to 7,600 A, 400 to 760 nm, .40 to .76 μm. Planck's equation tells us that a photon of blue light (400 nm) has almost twice as much energy as red light (760 nm).

Most light-measuring devices read the output of an electrical circuit. How is light converted to electrical units? In one such measuring instrument, light of a specific wavelength, focused onto a thin strip of platinum, raises the temperature of the platinum. This rise in temperature produces a rise of electrical resistance that can be measured by placing the platinum strip in a sensitive Wheatstone bridge circuit. A change of 1/1,000,000 in electrical resistance corresponds to a temperature rise of 1/4,000° C.[4] The watt (W) is the unit of power expressing work per unit of time.

Since watts are related to electrical resistance through Ohm's law,[‡] the device actually converts light via heat into watts. Conversely, when an electrical current flows through the filament, its resistance causes heat to be generated. The heat generated

---

[‡] Ohm's Law: Power (in watts) = $\dfrac{\text{voltage}^2}{\text{resistance (ohms)}}$

results in the filament's glow, producing visible radiation along with invisible radiation. An incandescent lamp rated 150 or 200 watts consumes the energy represented by its wattage rating. Its light output—with which we will be concerned in the next section, *Photometry*—is measured in lumens. The laser (Light Amplification by Simulated Emission of Radiation) is a source that often has its output in the visible wavelengths but has its output measured in watts.

The typical argon laser delivers up to 2 watts of power. Here we are talking about output watts—not the watts of power going from wall outlet to the laser. To calibrate, the laser beam is focused to the smallest spot possible onto a radiometer. The radiometer converts the light into watts for that tiny focused area. When the clinician uses large spot sizes, the amount of energy is simply spread out over the larger area, yielding a reduced power density.

The YAG laser creates an extremely high temperature in a very small volume in a very small time interval. Specifically, the temperature can rise to more than $15,000°\,C$ in the course of a pulse from a mode-locked or Q-switched YAG laser shot which lasts only nanoseconds (10^{-9} seconds) or picoseconds (10^{-12} seconds). This tiny explosion produces a mechanical rupture in tissue. Clearly, work is done in a very short time. Thus the work-term joules (a term without time) is an appropriate unit. For example, 5 to 50 mJ/10 micrometer, diameter spot size per burst, would be a typical YAG burst.[8]

Radiant Source

The sun is a radiant source. The best way to determine the intensity of a source is to measure the energy within a solid angle called a steradian. A steradian is defined as a solid angle emanating from the center of a sphere and subtending an area on the surface of the sphere equal to the square of the radius of the sphere. A sphere contains 12.57 steradians. Thus, the best way to describe the radiant intensity of a source is by the power that emanates, i.e., watts/steradian. From our vantage point on earth, we receive a minuscule portion of the sun's energy, i.e., a solid angle of about 5.8×10^{-9} steradians. To calculate the solid angle, one divides the area of earth by the distance of the sun to the earth, squared. This equals 5×10^{7} square miles divided by 93 million miles squared. Were we to convert the radiant power to radiant energy reaching the earth, it would be equal to 17 trillion trillion watts per steradian.

To measure the output of a surgical lamp, a radiometer is used to capture a small area of the output. Calculations that include the distance of the radiometer from the lamp and the receiving area of the radiometer are then made to arrive at the integrated output of the lamp, i.e., its radiant intensity in watts/steradian. Of course, the total radiant power, the *total* radiant flux included in the complete sphere, would be 12.57 times the watt/steradian (if the radiance were the same in all directions) and would be described in watts.

Radiation on a Surface (*Irradiance*)

As radiation diverges from its source, the density of energy falling on a surface decreases as the surface is moved farther from the source. It is known that density diminishes inversely as the square of the distance from the source. Thus, in terms of power/area, one could say that we on earth receive 73.3 W/sq ft or about 0.1 W/cm^2 of radiant energy from the sun.

To determine the amount of reflection from a surface, we multiply the irradiance by the surface reflection coefficient for a particular direction. For example, fresh snow reflects 85% of the incident solar radiation. Thus, one can get a slightly stronger suntan by facing the sun rather than facing the snow.

As noted before, units on the power meter for clinical argon and krypton lasers are in watts and milliwatts (0.001 W). Of course, such units by themselves do not describe the energy reaching the retina. The actual thermal lesion is dependent not only on source intensity, but also on the image size on the retina, the exposure duration, the clarity of the ocular media, and the absorption level for pigmentation. For example, the suggested minimal power level for panretinal photocoagulation in diabetic retinopathy might be 200 milliwatts, with a spot size of 200 microns and an exposure duration of 0.1 seconds. Of course, to tune the influence of media clarity and pigment absorption to the individual case, one usually increases the power setting from such a suggested starting level until a visible lesion is produced.

To safely obliterate new vessels near the macula in cases of senile macular degeneration, a smaller spot size such as 100 microns would be used along with a higher power setting of 300 to 600 milliwatts and shorter exposure (0.05–0.1 second). To produce an iridectomy in brown iris tissue, higher intensity levels (1,000 milliwatts) and longer durations (1 second) condensed into a 100 micron spot size are used.[3]

Retinal damage is not just caused by inducing thermal damage through short intense light exposure of lasers or the sun. Prolonged moderate light levels do not raise the retinal temperature but do produce photochemical damage to the photoreceptors in animals and possibly in man. We assume that this damage is usually repaired in man. The units of energy for this work are described as irradiance or power density, i.e., watts per square centimeter of eye tissue. For example, a 1-hour exposure of 1 milliwatt/cm^2 of blue light caused a functional blue blindness in monkeys,[9] whereas three quarters of an hour exposure of 7.4 milliwatts/cm^2 of white light produced electron-microscope-determined damage to a human subject's [§] rods and cones.[10] The many animal studies in which this phenomenon occurs are recorded in an excellent review by Lanun.[11]

Since photochemical retinal damage is so strongly time dependent, the clinician is advised to complete his work as quickly as possible when using (a) the indirect ophthalmoscope, which delivers a retinal illuminance of about 70 milliwatts/cm^2,[1] (b) the slit lamp, which delivers a retinal illuminance of 217 milliwatts/cm^2,[1] (c) the direct ophthalmoscope, which delivers a retinal illuminance of 30 milliwatts/cm^2,[1] and (d) a typical operating microscope, which delivers a retinal illuminance of about 1,000 milliwatts/cm^2.[11,12]

Psychophysics: Light Used as Illumination (Photometry)

Measurement of light connected to judgment of the sensation resulting from the light belongs in the realm of psychophysics. Photometric tests can never have the accuracy of strictly physical determinations. All attempts to eliminate the judgment of the human eye from photometry by replacing it with instruments have a built-in weakness that must be appreciated, for they are comparisons of power and not of physiological effect.

Although the intensity of radiation varies inversely with the square of the distance, the impression made on our eyes is approximately proportional to the logarithm of the distance. If colored light is considered, the physiological effect varies faster for long wavelengths (a higher power of the distance) and slower for shorter wavelengths (a lesser power of the distance).

Psychophysics relates stimuli with resulting sensations, but sensation is not perception. Sensation is a physiological result prompted by the stimulus. Perception requires that one call on memory and experience in a search for meaning. Perception, from the Latin *percipio*, means to sense wholly or completely. Perception evolves from the organizing functions of the entire nervous

---

[§] Eye of patient about to be enucleated for malignant meloma.

system, including the mind; and mind is used to signify in the broadest sense the *we* who think and will, and specifically not any single organ but rather the whole living organism.

Light Source in Photometry (the Illuminant)

The measurement of a light source started with an arbitrary but logical unit. A candle of certain thickness that burns up a certain amount of wax in a certain time was chosen as the standard, rated 1 candlepower. Such a candle was maintained in our Bureau of Standards and used as a reference. Today, we use a much more scientifically rigorous reference source, the *candela*. Please note all the qualifications in the definition. A candela is the luminous intensity in a given direction of a source that emits monochromatic radiation of a frequency 540×10^{12} Hz and of which the radiant intensity in that direction is 1/683 W/steradian.

As opposed to the traditional unit, the candle, the new definition for intensity of a source in a given direction, uses the physical term watt, and attempts to free itself from the spectral sensitivity of the eye by using one wavelength. However, by becoming more theoretically rigorous, it becomes an unreal abstraction. By freeing itself from white light, the candela may be more reproducible but it loses meaning in human terms. Then, too, none of the usual light sources have the spectral distribution of the radiator implied in the definition. We must simply appreciate that sources of light may be calibrated in candelas. Older literature will talk of candlepower. For practical purposes, 1 candlepower is about equivalent to 1 candela in luminous intensity. However, as mentioned in the section on physical radiation, the description of a light source is more meaningful if the concept of the integrated intensity light flux is used. Thus, most books and papers describing light sources will use units of light "flux," *lumens*. A common household 100-watt lamp produces approximately 1750 lumens when first put in service.

Light Flux of the Illuminant

The light flux from a light source is its total luminous output measured in lumens. The unit lumen (light in Latin is lumen) is the luminous flux from a source of 1 cd intensity contained in 1 steradian. As noted in the radiation section, the total light flux from a source of 1 cd in all directions (total light flux in a spherical envelope) would be 12.57 lumens.

Since our incandescent bulbs are rated in watts, let us take a sampling of the lumen output of some typical bulbs. It is interesting to note that as the filament goes from red hot to white hot, less heat and more visible light is produced. In fact, the theoretical maximum of a perfect source would be 94 lumens per watt of white light. However, the hotter the filament gets, the greater the chances of melting and the shorter the lifetime of the filament. The search for a filament material that would sustain a high temperature for many hours without melting led to the use of the tungsten filament in 1907, which could be operated at temperatures far higher than its predecessor, the carbon filament. The general service 100 W tungsten filament bulb of today puts out 1,750 lumens (17.5 lumens/W) and has a rated lifetime of 750 hours. The filament of a slide projector lamp rated at 100 W will get hotter, will put out 1,950 lumens, but will only last 50 hours. Bulb efficiency may also be improved by increasing the wattage. A 500 W bulb will put out 21.2 lumens/W, and a 1,000 W bulb can put out 23.6 lumens/W. The fluorescent lamp is more efficient producing 3½ times as much visible light per watt as the incandescent lamp, while developing only one half as much radiated infra-red energy and heat. A 40 W fluorescent lamp will produce about 2,100 lumens.

Illumination of a Surface (Illuminance)

The density of the light flux falling on a surface area is called the illuminance. This term must not be confused with luminance, which refers to the light reflected or emitted from a surface. Illuminance is measured in lumens per square centimeter, lumens per square meter, or lumens per square foot. For example, it has been suggested that a visual acuity chart [13] be illuminated by 10 to 15 lumens/sq ft or 10 to 15 foot-candles of illumination. A well-lighted office will provide 30 ft-c for most work surfaces. However, someone doing very fine work will require a personal lamp, because fine work may require 100 to 300 ft-c for efficiency. If we turn to the sun for another example, we note that light falling on the earth's surface is about 10,000 ft-c on a bright day; about 1,000 ft-c on a dull, cloudy day; and about 0.01 ft-c at night [14] (Fig. 1.1A).

Light Reflected from a Surface (Luminance)

Illuminated surfaces become light sources by either reflection (desk surface, paper, snow, etc) or transmission (translucent glass, etc). The term luminance (formerly called brightness) describes the luminous intensity of a surface. It relates to the intensity of light in a given direction (recall surfaces are often uneven). One may

A
B

Figure 1.1. Two pictures of the Anglo-Australian observatory dome taken 13 hours apart, one (A) by sunlight (left) and the other (B) by the light of a full moon. Both were taken with the same roll of ASA 400 color film, one with an exposure of 0.002 seconds at $f/22$, the other 11 minutes at $f/4$, so that the difference in exposure factor was 10 million. Both photos are essentially equivalent. (Courtesy of David Malin on behalf of the Anglo-Australian Telescope Board, copyright owner.)

regard reflected light as the expansion in size of a point source, becoming an area measured by its projection onto a plane at right angles to the viewing direction.

Luminance is the most complex of the photometric units and may also be the most perceptually misleading. For example, a luminance may differ because a surface only reflects, or because it both reflects and transmits. A reflecting surface may be specular, diffuse, somewhere between the two, or a combination of varying proportions of both. If one adds color to a surface then the luminance becomes more complicated.

A specular surface reacts directly to position of the source and direction of the ray; it is also sensitive to distance. Imagine yourself standing outside a window at sunset. If you can see the sun reflected in the glass, it occupies a small portion of the surface, and your vision may penetrate the glass and see some surrounding luminance of the room inside as contrast. Now imagine you are a mile or so away from the window, and it catches the setting sun. The entire window is ablaze with light. If you watch it for a few more minutes, it suddenly becomes dark, the sun having dropped below the horizon. If one alters his viewing direction, the reflected image changes. With specular surfaces the geometry involving the source, the surface, and the viewer all interrelate. This kind of detail is not a part of luminance calculations. Because the image

Table 1.2. Photometric terms.

Radius	1 centimeter	1 foot	1 meter
Area	1 cm^2	1 ft^2	1 m^2
Illuminance	1 *phot* or	1 *foot candle*	1 *lux* or
	1 lumen/cm^2	1 lumen/ft^2	1 lumen/m^2
Luminance	1 *lambert* or	1 *foot-lambert*	1 *apostilb*
	1 cd/cm$^2 \cdot \pi$	1 cd/ft$^2 \cdot \pi$	1 cd/m$^2 \cdot \pi$
Luminous emittance	1 lumen/cm^2	1 lumen/ft^2	1 lumen/m^2

Note: The use of lambert, foot-lambert, and apostilb is discouraged in illuminating engineering circles in favor of the descriptive identifications, which follow each of these terms. Other terms will be found occasionally and a proper text should then be consulted.

in a specular mirror is clearly detailed, a proper reading of all the luminances would constitute a numerical half-tone image of the reflected scene.

A perfectly diffusing surface reflects all incident illumination and its directional intensity is well represented by its luminance. No surface is perfectly diffusing and all of the intermediate conditions between specular and diffuse along with their combined characteristics give rise to such a variety of luminance conditions that the single number metric fails to convey the differences adequately. All of these observations are equally true of transmitting media.

If a square foot of diffusing surface reflects 1 lumen of light we can say that the surface has a luminance of 1 foot-lambert. Units of illuminance and luminance are to be found in Table 1.2. For example, the bright sky on a clear day gives a background luminance of 1,000 foot-lamberts, whereas a sunlit cloud gives 10,000 foot-lamberts. The ground on a sunny day would reflect 100 foot-lamberts, whereas snow in full sunlight would reflect 5,000 foot-lamberts. On the other hand, foot-lamberts are also used to describe dark backgrounds. Thus, a dark overcast sky with no moon yields 0.00001 foot-lamberts whereas a clear moonlit sky yields 0.01 foot-lamberts (Fig. 1.1B).[14]

Psychological Terms: Visual Perception

Turning now to brightness, lightness, and color is to enter the psychological world of perception where all properties are *relative*. We have left the physical world of the absolute and dwelt for a considerable time in the physiological realm of the psychophysical, where external physical changes are coupled with sensations, forming the "laws" of photometry. We now enter the psychological world. Here the individual relates on a one-to-one basis with his surroundings and with other people. The terms are qualitative rather than quantitative. Objects and events are described as "more" or "less," but how much more or how much less cannot be given any meaningful metric. John Ruskin warned the painter,

Every hue throughout your work is altered by every touch that you add in other places; so that what was warm a minute ago, becomes cold when you have put a hotter color in another place, and what was in harmony when you left it, becomes discordant as you set other colors beside it.[4]

To recognize the limits in the application of photometric terms, it is revealing to compare them with their counterparts in the glossary of the artist. Albert Munsell devised a system of nomenclature to describe object properties. His terms are *hue*, for color perception as red, orange, yellow, etc.; *value*, which describes the lightness, the gradation from black to white; and *chroma*, the amount of color or lack of it from full color to gray. His work is especially valuable since the steps in each of the three scales are equal perceptual steps. The numbers are determined by the psychophysical process of establishing least perceptible changes or by partitioning (dividing in half, then in half again, and so on). Deane B. Judd and Gunter Wyszecki find that "the Munsell color terms correspond closely with the attributes of the color perception itself."[14]

Brightness is missing from Munsell terminology and may be the most difficult concept to formalize. *Brightness* is the subjective attribute of any light sensation giving rise to the percept of luminous magnitude. Its scale ranges from blackness or invisibility, arising from the absence of light, to dazzling, the maximum sensation our eyes can give us. *Lightness* is the property by which a surface is perceived to transmit or reflect a greater or lesser fraction of the incident light.

For lightness of color Munsell gave the name *value* and to it he assigned ten perceptually equal steps between black and white. Although these steps appear equal, their reflectance intervals vary widely. A 0% reflectance is black; 19.3% is midway between black and white, designated as 5 in the Munsell scale; 57.6% reflectance for Munsell value 8; and 100% for white. (There are no reflectors having either 0% or 100% values.)

The complete Munsell system identifies more than 1,500 colors, and there are color samples available for each of them. The only certain way of communicating a complete idea of a color is by means of a sample.

The conditions under which the steps of the scales are seen to be equal are carefully stated: a normal observer fully adapted to daylight viewing conditions, with a gray to white surround. The importance of the conditions is what is left unspoken; they, the carefully stated conditions, rarely prevail. Under other circumstances the steps may become unequal. Remember that no single individual may qualify as a "normal" observer.

Rudolph Arnheim noted, ". . . we have trained ourselves to rely on knowledge rather than our sense of sight to such an extent that it takes accounts by the naive and the artists to make us realize what we see." He added:

The color of most objects is anything but uniform in space or time; nor is it identical in different specimens of the same group of things. The color the child gives to the trees in the picture is hardly a specific shade of green selected from the hundreds of hues to be found in trees. Again we are dealing not with an imitation but with an invention, the discovery of an equivalent that represents the relevant features of the model with the resources of a particular medium.[15]

Summary

We have tried to demonstrate that light must be divided into light as energy (a purely physical phenomenon) and light for seeing. Light for seeing can be usefully measured in psychophysical terms. However, such terms do not represent the interplay of mind and brain on the psychophysical stimuli. One does not have to be an artist to appreciate the richness of visual perception. Each category—light as radiation, light as a psychophysical stimulus, and light as a visual perception—has its own set of terms and measurements. It is hoped that this chapter will help you to know which situation requires which terms and to know what the terms mean.

References

1. Calkins JL, Hochheimer BF: Retinal light exposure from ophthalmoscopes, slit lamps, and overhead surgical lamps. An analysis of potential hazards. Invest Ophthalmol 19:1009, 1980.
2. Lesperance FA: Ophthalmic Lasers, Photocoagulation, Photoradiation and Surgery. CV Mosby, St Louis, 1983.
3. Sykes SM, Robinson WD, Waxler M, Kuwabara T: Damage to the monkey retina by broad spectrum fluorescent light. Invest Ophthalmol 20:425, 1981.
4. Erhardt L: Radiation, Light and Illumination: A Re-creation. Camarillo Reproduction Center, Camarillo, CA, 1977.

5. Moon P: The Scientific Basis of Illuminating Engineering. McGraw-Hill Book Company, New York, 1936.

6. Sliney D, Wolbarsht M: Safety with Lasers and Other Optical Sources. Plenum Press, New York, 1980.

7. Steinmetz CP: Radiation, Light and Illumination. McGraw-Hill Inc, New York, 1909.

8. Trokel SL: YAG Laser Ophthalmic Microsurgery. Appleton-Century-Crofts/Prentice Hall Inc, East Norwalk, CT, 1983.

9. Sperling HG, Harwerth RS: Intense spectral light effects on spectral sensitivity. Opt Acta 19:395, 1972.

10. Radnot M, Jabbagy P, Heszeiger I, Lovas B: Les données à l'ultra structure de la rétine humaine sous l'effet de la lumière. Ophthalmologica 159:460, 1969.

11. Lanun J: The damaging effect of light on the retina. Empirical findings, theoretical and practical implications. Surv Ophthalmol 22:221, 1978.

12. Parver LM, Auker CR, Fine BS: Observations on monkey eyes exposed to light from an operating microscope. Ophthalmology 90:964, 1983.

13. Borish IM: Clinical Refraction. The Professional Press, 1970, pp 357, 358.

14. Judd DB, Wyszecki G: Color in Business, Science and Industry. John Wiley & Sons Inc, New York, 1975.

15. Arnheim R: Art and Visual Perception. University of California Press, 1954.

2

The Photochemistry of Life and Cell Death: A Philosophical Overview

David Miller

Introduction

I suppose that I've always known the principle, and yet it came as a surprise when it was laid out plainly before me. In the past, when I've prescribed a potent drug, corticosteroids, or cytotoxic agents, I've always warned the patient that these drugs can act as two-edged swords. Alfred Nobel taught us that dynamic could both build a civilization or level one. Yes, we all know that drugs and chemicals do harm as well as good. But it was a shock to realize that such fundamental entities as light and oxygen fall into the same category, and even more so. Sunlight and oxygen, along with natural pigments, not only are the cause of all life on this planet but can also be responsible for aging and death.

For the moment, let us turn to the question of the origin of life on earth and review some of the fascinating details.

Origin of Life

More than 100 years ago du Bois Reymond recognized that the more he learned about the science of physics, the more the beginnings of life on earth seemed impossible. He had realized that the random Brownian movement of molecules was too weak to form new organic compounds. He felt that only the short wavelengths of ultraviolet (UV) light were energetic enough to produce the ternary and quaternary organic compounds from primordial carbon dioxide, steam, and ammonia. He did not speculate whether UV rays came directly from the sun or were the result of lightning bolts. However, the type of energy level that he was speculating about are those found in the short wavelength UV light produced by our modern-day germicidal lamps. Isn't that odd? The very light needed to create life destroys it. The story becomes even more confusing. Our present atmosphere does not allow these high-energy UV rays to strike the earth. In fact, only 5% of all light striking our planet is in the UV range. Of this UV light, 97% is UV-A.[*] If you think about it for awhile, a possible set of clues emerges. Our present atmosphere may be different from the original one. We know that the atmospheres of Venus and Mars contain no oxygen. Presumably our primitive atmosphere was similar to their present one. Such an atmosphere would allow short UV waves to strike the earth. Is there any way to allow UV rays to create life and not turn around and destroy it? We know that a thin layer of seawater absorbs UV rays below 280 nm, and a 1-m-thick layer will absorb below 320 nm. Of course, murky water is a more efficient absorber and scatterer. Thus, a thin layer of murky water could titrate out the more potent UV light, allowing the friendlier, longer UV and visible radiation to penetrate. The key variable in this argument becomes water depth. Perhaps the first "Adam" bacteria developed just below the surface of the primordial sea and not within the atmospheric vapors after a bolt of lightning. Thus, life had to wait until the original earth cooled down to a temperature at which the atmospheric steam condensed into oceans. Since the atmosphere had no oxygen, our earliest ancestors were obligate photosynthetic anaerobes. It is estimated that this occurred in the Precambrian era some 3.5 billion years ago [1,2†] (Fig. 2.1). Information about the next evolutionary stages is vague, but we can surmise that a series of oxygen-producing organisms started to develop. Geologic layering suggests that the first evidence of the presence of oxygen shows up in red rock containing ferric oxide. Such layers occurred in the Precam-

---

[*] Ultraviolet light is broken down into UV-A (320–400 nm), UV-B (290–320 nm), and UV-C (200–290 nm). Doses of UV-B radiation is 100 to 1,000 times more damaging to tissue than UV-A. Most of the sun's radiation is in the form of infrared. About a third of the sun's output is in that visible range.

[†] The oldest fossil record indicating any kind of life on earth is found in rocks at least 3.5 billion years old. These primitive bacteria forms were the only living creatures for the next 2 billion years. The planet is estimated to be 4.6 billion years old. These "Adam" bacteria (known as purple sulfur bacteria today) contain a chlorophyll that absorbs a wider spectrum of visible light than plant chlorophyll.

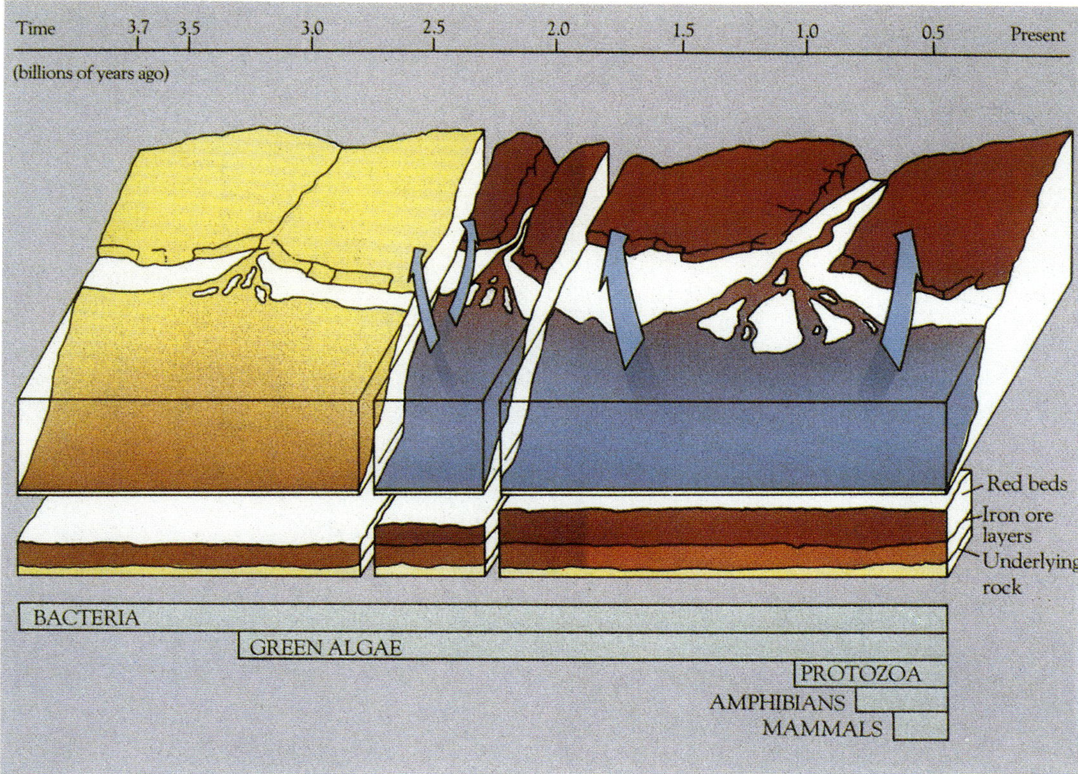

Figure 2.1. Three-dimensional scale illustrating time on earth v evolution of different species v the development of different earth and atmospheric layers. (From *The Intelligent Universe* by Fred Hoyle. Copyright © 1983 by Dorling Kindersley Limited. Reprinted by permission of Henry Holt and Company.)

brian era some 500 million years ago.[1] Oxygen produced by these early organisms not only dissolved in the ocean, but rose in the atmosphere,[‡] reacting with the UV light to produce ozone (Fig. 2.2). Ironically, as the ozone layer built up some 20 to 30 km above the earth, it prevented the short wavelengths from getting through.[§] Thus, a period of frenzied evolution must have taken place, in which photosynthetic pigment kept developing and ab-

---

[‡] We tend to take our atmosphere for granted. Atmospheres only exist around stars and planets with significant gravitational fields. Thus the moon is too small to have an atmosphere. The earth, on the other hand, is heavy enough to attract the gases of our atmosphere.

[§] Under standard conditions, the ozone layer varies between a thickness of 2.4 mm to 4.6 mm. This variation depends on the season, latitude, and sunspot activity.

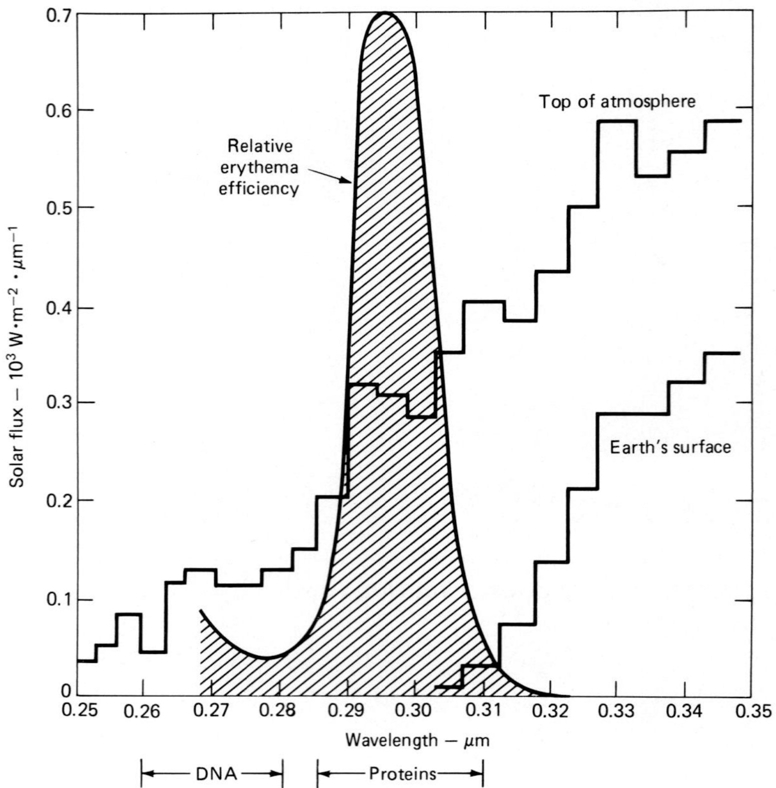

Figure 2.2. Graphic representation of how the ozone layer absorbs a segment of the sun's UV radiations, with a peak at 290 nm. Note how little UV below 300 nm strikes the earth after passing through the ozone layer. (From Mac Cracken M, Chang J: Preliminary study of the potential chemical and climatic effects of atmospheric nuclear explosion. UCRL 51653, Lawrence Livermore Lab, April 25, 1975, p 48. Courtesy of University of California, Lawrence Livermore National Laboratory and the Department of Energy.)

sorbing longer and longer wavelengths as the ozone kept cutting out the shorter ones. This wild play of molecular agitation, valence indecision, and laws of photochemistry ultimately resulted in the appearance of blue-green algae containing chlorophyll.

The photosynthetic production of oxygen by this tiny creature not only slammed the door on the lethal UV rays, i.e., those below 290 nm, but also set the stage for the more energy-efficient aerobic glycolysis. This cycle would ultimately power multicellular organisms.

Although the phenomenom of photosynthesis may be 500 million years old, its story is of recent origin. Daniel Arnon, himself a major modern contributor to the understanding of photosynthesis, has put the story together in a most charming way.[3] He notes that the first peek into the process took place in 1771 when a clergyman-scientist by the name of Joseph Priestly made an unusual observation.

Priestly described how he had "injured the air" inside an inverted glass jar by burning a candle. He then was able to "restore the air" by allowing a sprig of mint to grow in a vessel of water under the glass jar. As proof of the restoration of the air, he placed a burning candle and a mouse under the glass jar and observed that "the air would not extinguish the candle nor was it at all inconvenient to the mouse." Thus Priestly concluded that "plants were able to keep the atmosphere sweet and wholesome." As Arnon noted, Priestly's findings provided a rational theory for planetary ventilation, explaining how plants have kept the earth's atmosphere healthful (despite the continued breathing of humans and animals for thousands of years). But a strange thing happened; when others tried to repeat Priestly's experiments, they failed. As a matter of fact, Priestly himself could not repeat them.

The puzzle attracted a Dutch physician named Jan Ingenhousz who was working at the court in Vienna at the time. He was so drawn to this mystery that he obtained a leave of absence from the Queen and traveled to London so that he might get a closer look at all the clues.

He rented a small villa outside London and proceeded to set up a modest laboratory. Working night and day, it took him only 3 months to arrive at the solution. His own words best describe his findings:

> I observed that plants not only have a faculty to correct bad air in six to ten days by growing in it, as the experiment of Dr. Priestly indicates, but they perform this service in a complete manner in a few hours. This wonderful operation is by no means owing to the growth of the plant, but to the influence of the light of the sun upon the plant. This service is not performed by the whole plant but only by the leaves and green stalk that supports them.

Ingenhousz goes on to say, "When I recognized this astonishing difference between the effects produced by plants that receive light and those which are in darkness, I no longer had any difficulty in reconciling the variable, inconsistent and often contradictory results of the experiments of Drs. Priestly and M. Scheele." But what did the plant make "good air" from? Nine years later this piece of the puzzle was found by Jean Senebien, a Swiss Pastor. Using leaves submerged under water, he discovered that plants need "fixed air" and carbon dioxide to produce good air. Professor

Arnon notes, "the chemical balance sheet of photosynthesis was finally completed in Geneva by Nicholas Theodore de Saussure who found that water was also a reactant in the process." Thus by the turn of the 19th century, photosynthesis was seen as a reaction occurring in the green portions of the plant, under the influence of sunlight, in which CO_2 and H_2O were converted into carbohydrate and oxygen. Today, we know that even more photosynthesis takes place within our oceans than on land.

In 1845 Julius Rober Mayer,[1] a German physician who had formulated the law of the conservation of energy, refocused our attention on photosynthesis. "Nature set herself the task of capturing the light flooding toward earth and storing this. . . . The plant world constitutes a reservoir in which the fleeting sun rays are fixed and ingeniously stored for future . . ." How prophetic Mayer was. Today, modern man runs his factories and cars with the energy stored in the fossils of ancient organisms and plants that were buried 100 to 400 million years ago.

As our technology advanced, scientists were able to probe more deeply into the more minute workings of the little plant solar cell.

On the one hand they saw the noonday sun radiating 1,300 J/m^2 of energy every second.[4] This is the result of billions of tiny thermonuclear reactions in which four hydrogen nuclei combine to form one helium atom.[3] The differences between the mass of the helium nucleus and that of the hydrogen is the energy that keeps the earth's temperature at the proper setting and fuels photosynthesis.|| On the other hand they watched the plant generate and store chemical energy. The plant's energy was packed in the energy-rich phosphate bonds of the molecule adenosine triphosphate (ATP). This molecule, so essential to life (its presence actually defines life for the biochemist), also drives the vital processes of all bacterial and animal cells.

A link was missing. Exactly how was solar energy converted into chemical energy? With the advent of the new technique of electron-spin resonance spectroscopy and flash photolysis, the events of the first 100 billionth of a second of photosynthesis have been defined. For this work, a chlorophyll-containing bacterium was used as the subject. It was learned that upon absorbing four

---

|| It is estimated that inside the sun, about 564 million tons of hydrogen are converted to 560 million tons of helium every second. The bottom line result is the conversion of 4 million tons of matter into radiant energy. Only a fraction of one millionth of 1% of this energy is used to sustain life on our planet. These reactions are expected to continue for about 5 billion years.[30,31]

quanta of red light, the chlorophyll molecule reaches a higher
energy level and becomes known as a free radical.[5]

Free Radicals

Light has provided an excited electron. Life must catch the electron in its excited state, uncouple it, and lead it back to a ground
state through a factory of enzyme machinery. In this case the
electron will return to the chlorophyll molecule after helping to
form ATP. Once in a chemical form, the energy will work to
break down water and carbon dioxide. The energy left over will
be stored in the resulting carbohydrates and fats of the plant.

For a moment it might be useful to stop and appreciate the
enormity of this process. It has been calculated that all the life
on our planet—plankton, bacteria, insects, plants, animals, etc—
occupies about 95% of the earth's surface, numbers over 10^{31}
living creatures, and weighs about 2×10^{19} g.[1]

In a sense, one may say that this overwhelming amount of life
was started and is maintained by the energy of sunlight, through
the agency of the photochemical creation of free radicals.

Let's look at the free radical in more detail. If we think back
to our basic courses in chemistry, we remember that the flame
of the bunsen burner was necessary to wrench open bonds and
force the molecular collisions that produced the new compounds.
However, the reaction we are talking about is much more subtle.
Chlorophyll is a rather sophisticated molecule with its four hydro-
carbon pyrrole rings surrounding a couple of magnesium atoms
(actually a close relative to hemoglobin). Such a configuration main-
tains spacings that can accommodate quanta of specific wavelengths. One such wavelength is 690 nm, i.e., red light (Fig.
3.3, 3.4). According to quantrum physics, these quanta are less
than half as energetic as those that produced the Adam or Eve
bacterium. However, the miracle of chlorophyll is that it can work
with visible light and still get the job done. The presence of these
photons or quanta of red light breaks up a pair of electrons living
within the molecule and not involved in actual bonding. One of
these electrons is driven into a new orbital where its spin is not
neutralized by a partner. Thus its magnetic moment is not neutralized and the molecule takes on a paramagnetic quality. This new
state is referred to as either a triplet or singlet state. The singlet
state falls back to normal in only a fraction of a second whereas
the triplet state may take up to a full second to revert back. This
momentary rearrangement of the "furniture and fittings" of the
molecule results in redistribution of the charges all over. The

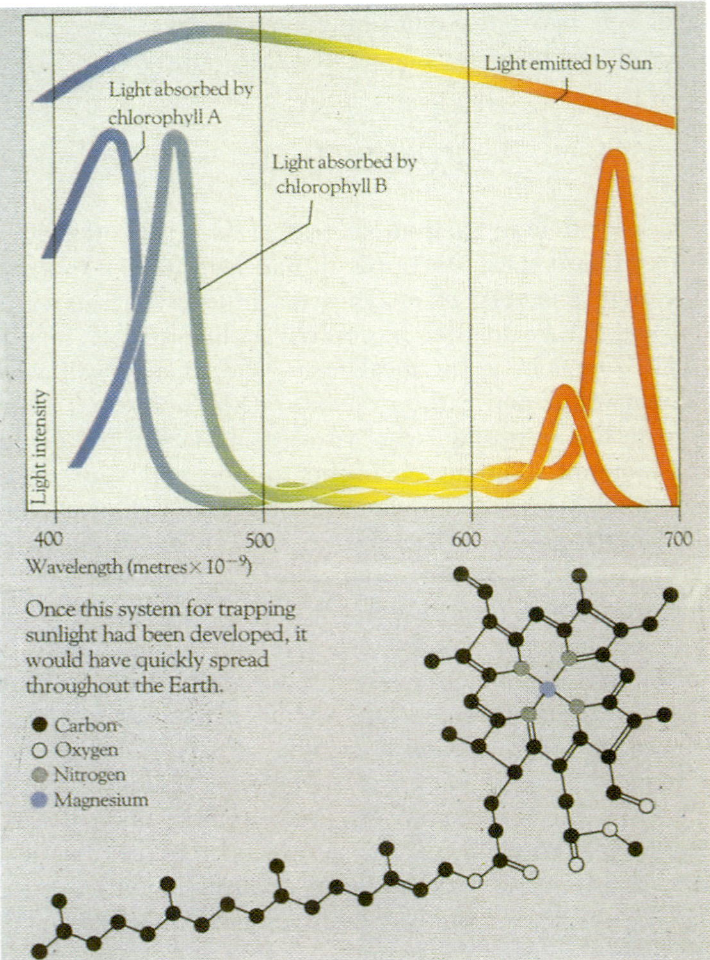

Figure 2.3. Absorption spectrum and molecular configuration of chlorophyll. Chlorophyll absorbs maximally in the blue and red regions of the light spectra and reflects in the green. (From *The Intelligent Universe* by Fred Hoyle. Copyright © 1983 by Dorling Kindersley Limited. Reprinted by permission of Henry Holt and Company.)

molecule is now in a frenzy, i.e., a higher energy state. If we take a close look, we can see that the C-C bonds in the ring structures widen and become weaker.[6] The laws of the economy of nature demand that the excited molecule, now known as a free radical, must return to its ground state. It may do this by either spitting out a photon, i.e., it may fluoresce, or it may transfer the transplanted electron to a neighborhing molecule or, being more reactive, it may combine with a neighbor to form a new compound. Alex Comfort, author of *The Joy of Sex*, once said, "A free radical has been likened to a convention delegate away from his wife: it is a highly reactive chemical agent that will com-

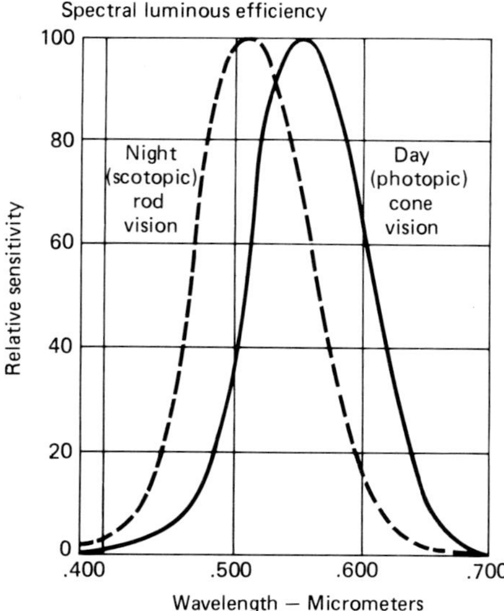

Figure 2.4. Spectral sensitivity of the light-adapted and dark-adapted eye. It is interesting to note that the eye's photopigments are maximally sensitive in the green. In a sense, the reflected wavelengths of chlorophyll fit nicely into the optimal wavelengths processed by the eye.

bine with anything around." For the chlorophyll-free radical, the convention lasts only a fraction of a second. However, there are other free radicals produced by light which last longer and have more explosive effects. Although the superoxide radical will be discussed in great detail in later chapters, the story of its discovery would be appropriate here.

It should be clear by now that a deep understanding of the free radical involves calculations of bond strengths, bond lengths, molecular energy levels, and many other variables. In 1930 it was thought that molecular oxygen only existed as O_2. At that time, Linus Pauling did some calculations and predicted that oxygen could also exist as the radical, superoxide (O_2^-), i.e., carry an extra electron.[7] He asked his student Edward Newman to prepare potassium oxide and in the process search for the presence of the superoxide radical. Sure enough, Pauling was right and the superoxide radical became an accepted entity. As one might expect, such an excited species can become even more reactive and ultimately cause damage. It might combine with the molecules of cell membranes or even react with DNA itself, causing mutations or death. Therefore, we have come back to the thought that started this chapter. There is a chemical species that is capable of initiating both life and death.

Let us now take a closer look at this sinister side of the free radical.

Photodynamic Action

This story begins during the winter of 1897. A young medical student named Oscar Raab gained a job in the laboratory of Professor von Tappeiner at the University of Munich. The professor had observed that the dye acridine orange was capable of destroying the one-celled paramecia, but the effect was inconsistent. In time, the diligent Raab learned that the dye alone did not kill the organism. The dye only had its destructive effect in the presence of light.[8] One wonders if young Raab knew of Ingenouzs' work on photosynthesis and simply applied the same principle. In any event, the student's work stimulated von Tappeiner, and in time other dyes such as eosin and chlorophyll were shown to destroy many different organisms in the presence of light. As the work progressed, von Tappeiner's group learned that both oxygen and light were needed if the dye was to kill the organism. Von Tappeiner coined the term photodynamic action to describe the special killing effect of a dye under the influence of light and oxygen.[9] Today, we call this group of dye and pigments, photosensitizers. The group includes dyes such as fluorescein, rose bengal, methylene blue, and acridine orange; natural pigments like chlorophyll and bilirubin; and drugs such as tetracycline. Their general chemical structure almost always includes a number of closed rings and alternating double bonds.[10] This molecular configuration is able to absorb UV photons. Such photon trapping promotes the molecule to a higher energy state. If molecular oxygen O_2 is close by, a fresh electron is transferred to the oxygen, which is then said to have been reduced to the superoxide radical. In the proper environment, superoxide radical can link to hydrogen and form hydrogen peroxide. The hydrogen peroxide may gain a proton (H^+) and convert to the hydroxyl radical (OH^-, the most potent oxidant known.[11] This super villain has the reputation of reacting with and destroying the proteins and fat[12] of cell membranes, mitochondrial membranes, and nucleic acids.[13] Of course, the terms villain and hero simply reflect personal prejudice. Breaking down the cell membranes of virulent bacteria would also be compatible with goodness and virtue. For example, the creative Scandanavian physician Niels Ryberg Finsen was awarded the Nobel prize in 1903 for developing a system of using UV light [¶] to cure the painful skin lesions of tuberculosis.[14] One can only assume

---

[¶] Finsen and an engineer (afflicted with lupus vulgaris) designed an optical instrument using a carbon arc as a source of UV-B radiation and a lens made of a quartz flask filled with water. Such a system transmitted UV-B, but the water absorbed the heat.

that the germicidal role of the UV light was facilitated by the oxygen levels of the tissue and of a photosensitizer, present in either the skin or in the organism.

Although the emphasis of this chapter has been on the induction of free radicals by light energy, it should be mentioned that the superoxide radical and its hyperactive relatives, i.e., H_2O_2 and OH^-, can be formed in other ways. For example, active phagocytosis of leukocytes is associated with the generation of the superoxide radical. In fact, it is believed that the leukocyte kills the engulfed bacterium by immersing it into a pool of superoxide and hydroxyl radicals, which explode the organisms. It has also been suggested that the release of these radicals in the area of invasion is akin to pulling the alarm that sets off a cascade of inflammatory mediators.[15] At present, it is felt that the superoxide radical is generated by the leukocyte membrane-bound enzyme nicotinamide adenine dinucleotide phosphate, reduced form (NADPH) oxidase.[16] The enzyme xanthine oxidase with the substrate xanthine has also been shown to induce the superoxide radical in milk,[17] as have other enzymes. The superoxide radical has also been shown to be a by-product of normal cell respiration,[18] probably helping to generate ATP in a more efficient manner.[19]

Finally, the superoxide radical can also be produced by the auto-oxidation of the herbicide, paraquat, the liver toxin carbon tetrachloride, and the cancer cytotoxin, adriamycin.[20] Perhaps the release of O_2^- is their mechanism for killing?

Free-Radical Diseases

We have now arrived at a rather unsettling concept. The oxygen that we have grown up to believe supports our lives may also play a more sinister role. As mentioned above, it may be converted to the superoxide radical and produce damage in the parent organism. Actually, a number of degenerative and disease processes have already become associated with superoxide damage.[18] These include the following:

A. Ionizing radiation damage
B. Skin diseases
    skin cancer
    photoallergy
    solar urticaria
    skin lesions of lupus erythematosus, pemphigus, and bullous pemphigoid
C. Hyperoxygenation syndromes
    respiratory distress syndrome
    hyperbaric oxygenation damage

   D. Ischemia/reperfusion syndromes
       myocardial infarction
       cerebral ischemia
       intestinal ischemia
       shock
   E. Arthritis
   F. Aging

Later chapters discuss the role of free radicals in cataract, retrolental fibroplasia, senile macular degeneration, and other eye problems.

Free-Radical Scavengers

The oxygenation of our biosphere, which started with the Eve/Adam blue-green algae, imposed a number of evolutionary pressures. While the organisms worked on developing more efficient methods of using oxygen, they all had to work out ways of coping with the toxicity of oxygen.

Actually, no group in science or industry is more familiar with the toxicity of oxygen than the food chemists. Everyday they must battle against the spoilage and discoloration of food. At the appropriate temperature, oxygen, under the sponsorship of either light or catalyst, can be transformed into the evil superoxide and degrade food. The most vulnerable targets are lipids (particularly unsaturated lipids), thus the high vulnerability of butter and milk to turn rancid. Therefore, the food chemist has learned to use antioxidants to preserve food. Popular antioxidants include vitamin C, vitamin E, butylated hydroxytoluene, butylated hydroxyanisole, and sodium benzoate.

Nature, too, has developed a system of disarming the superoxide radical. When biochemists looked for natural antioxidants, they found them in every nook and cranny of the body. For example, the blood contains at least two potent antioxidants: copper-containing ceruloplasmin and iron-transporting transferrin. both appear to prevent the ferric ion from giving up electrons to oxygen and thereby producing superoxide.[17] Another common scavenger found throughout the body is glutathione perioxidase, which you will read more about in later chapters. Many studies have shown that photosynthetic bacteria would be killed by photodynamic action were it not for the protective effect of carotene pigment.[21] However, the most potent and most pervasive scavenger is known as superoxide dismutase (SOD). For many years this extraordinary enzyme remained incognito, thought to be simply another copper storage enzyme. Everything changed in 1969 when McCord and Fridovich discovered that SOD was responsible for the rapid disarming of superoxide, wherever it was.[22] Its mechanism of action is much like the bomb squad; it simply removes the detonator cap, in this case that extra electron. Once the importance of the discovery of SOD's function was understood and appreciated, the critics started asking very pointed questions. If the superoxide radical is so dangerous and SOD is so effective in quenching the radical, shouldn't SOD be found in all oxygen-respiring cells, including all aerobic bacteria. The answer turned out to be essentially yes. Aerobic bacteria have a good supply of SOD or other potent scavengers on board at all times. Well then, do anaerobic bacteria die in the presence of oxygen because they have no SOD. With the exception of some odd but understandable exceptions, the answer was again yes.[23–25] As more information surfaces about O_2^- and SOD, they seem to gain greater and greater status as two of the fundamental forces involved in the yin and the yang of life. For example, in 1980 Tolmasoff et al[26] wondered if the ratio of SOD-specific activity to specific metabolic rate in an organism had a direct correlation with life span potential. Using two different rodents and 12 primate species, they accumulated a group of animals that represented a life span potential from 3.5 to 95 years. In analyzing the issue from the animals they discovered that the longer-lived species did indeed have a significantly higher degree of protection against the by-products of oxygen metabolism.

But what is considered the elemental biochemical factor in aging? Analysis of lung collagen in elderly people showed an increased number of cross-links.[27] Increased cross-linking and rigidity is also known to increase in the cell membrane of the elderly.[28] The pathologist knows that tissue is best fixed, i.e., strongly cross-linked, by formaldehyde or gluteraldehyde. Thus aldehydes are strong cross-linking agents. It has been shown that the superoxide radical can react with the double bonds of polyunsaturated fatty acids of tissues and membranes to produce lipid peroxidases.[20] Lipid peroxidases are notorious for decomposing into aldehydes. The aldehydes can then cross-link the proteins of enzymes, membranes, and collagen. The superoxide radical-induced peroxidase has also been shown to chemically break up and then congeal subcellular organelles like liposome, mitochondria, and microsomes. The result of these reactions is the accumulation of the aging pigment lipofuscin.[29] Lipofuscin granules often act like useless lumps of garbage, interfering with cell function. More will be said about the role of lipofuscin in the chapter on retinal disease.

In summary, we have discussed the relationship of certain wavelengths of light, along with pigments and oxygen-fueling life processes. We have also noted that other wavelengths, notably UV rays, along with pigments and oxygen, can combine to produce free radicals, which both support life and also cause cell damage and death. We have then talked about free-radical scavengers designed to protect the body from the harmful effects of potent free radicals such as the superoxide family. Clearly, all these factors are in proper balance when tissues are healthy. This book will discuss certain eye diseases that seem to develop when these factors go out of balance.

References

1. Terrestrial-Global Spectal Irradiance Tables for Air Mass 1.5″. ASTM Document 138 RI E44.02, Feb. 1981.
2. Dauvillier A: The Photochemical Origin of Life. Academic Press, New York, 1965.
3. Arnon DI: Sunlight, earthlife,: The Grand Design of Photosynthesis. The Sciences, October 1982, pp 22–27.
4. Kocherar IE, Anderson RR: Experimental Techniques in Photochemistry, in Photoimmunology, Parrish JA, Kripke ML, Morison WL (eds). Planum Medical Arts, New York, 1983, p 53.
5. Loach PA, Hales BJ: Free Radicals in Photosynthesis, in Free Radical in Biology, vol 1, Pryor WA (ed). Academic Press, New York, 1976.
6. Glass B, McElroy ED: A Symposium on Light and Life. Johns Hopkins Press, 1961 pp 817–830.
7. Pauling L: The Discovery of the Superoxide Radical. Trends in Biochemical Science, November 1979.
8. Raab O: Uber die Wirkung fluorescirender staffe auf Infusorien. Z Biol 39:524–546, 1900.
9. Von Tappeiner J, Jodlbauer A: Uber die Wirkung der photodynamischen (fluorescierenden) staffe auf Protozoen and Enzyme. Disch Arch Klin Med 39:427–487, 1904.
10. Spikes JD: Photodynamic reactions in photomedicine, in The Science of Photomedicine. Regan JD, Parrish JA (eds). Plenum Press, New York, pp 113–147, 1982.
11. Dorfman LM, Adams GE: Reactivity of hydroxyl radical aqueous solutions. NSRD-NBS, No 46, US Department of Commerce, National Bureau of Standards, 1973.
12. Mead JF: Free radical mechanisms of lipid damage and the consequences for cellular membranes in Free Radicals, in Biology. Pryor C (ed). vol 1, Academic Press, New York, pp 51–67, 1976.
13. Carrier WL, Snyder RD, Regan JD: Ultraviolet Induced damage and its repair in human DNA, in The Science of Photomedicine. Regan JD, Parrish JA (eds). Plenum Press, New York, pp 91–113, 1982.
14. Sourkes TL: Nobel Prize Winners in Medicine and Physiology: 1901–1965. Abebland-Schuman Inc, London, pp 15–19, 1966.
15. DelMaestro RF, Thaw HH, Bjork J, Arfoors KE: Free radicals as mediators of tissue injury, in Free Radicals in Medicine and Biology.

Lewis DH, DelMaestro RF (eds). Acta Physiol Scand [Suppl] 492:43–59, 1980.

16. Dewald B, Baggiolini M, Curnutte JT, Barbior BM: Subcellular localization of the superoxide forming system in neutrophils. J Clin Vest 63:21–29, 1979.

17. Kellog EW, Fridorich I: Superoxide, hydrogen perioxide and singlet oxygen in lipid peroxidation by the xanthine oxidase system. J Biol Chem 250:8812–8817, 1975.

18. Dormandy TL: Free radical oxidation and antioxidants, Lancet, March 25, pp 647–650, 1978.

19. Loschen G, Azzi A, Ilohe L: Mitochondial H_2O_2 formation: relationship with energy conservation. FEBS Lett 33:84–87, 1973.

20. Leibovitz BE, Siegel BV: Aspects of free radical reactions in biological systems: Aging. J Gerontol 35 (1):45–56, 1980.

21. Krinsky NI: Carotenoid protection against oxidation. Pure Appl Chem 51:649–660, 1979.

22. McCord JM, Fridovich I: An enzymatic function for erythrocuprein (hemocuprein). J Biol Chem 244:6049–6055, 1969.

23. Friedovich I: Oxygen radicals, hydrogen peroxide and toxicity in Free Radicals in Biology. Pryor WA (ed). Academic Press, New York, pp 239–271, 1976.

24. Friedovich I: The biology of oxygen radicals. Science 201:875–879, 1978.

25. McCord JM: The suproxide free radical: The biochemistry and photophysiology. surgery 94:412–414, 1983.

26. Tolmasoff JM, Ono T, Cutter RG: Superoxide dismutase: Correlation with lifespan and specific metabolic rate in primate species. Proc Natl Acad Sci USA 77 (5):2777–2781, 1980.

27. Rickert WS, Forbes WF: Changes in collagen with age. VI: Age and smoking related changes in human lung connective tissue. Exp Gerontol 11:89–101, 1976.

28. Nagy IZ: A membrane hypothesis of aging. J Theor Biol 75:189–195, 1978.

29. Chio KS, Reiss U, Fletcher B, Tappel AL: Peroxidation of subcellular organelles: Formation of lipofuscin like fluorescent pigments. Science 166:1535–1536, 1969.

30. Hawkes J: Man and the Sun. Random House, New York, 1962.

31. Menzel DH: Our Sun, rev ed. Harvard University Press, Cambridge, Mass, 1959.

Light Damage to the Eye

3

Perspective on Damage to Angle Structures

P. John Anderson and David L. Epstein

Light

The eye is an organ of light. The functional study of the tissues of the optic axis is long established. Our understanding of the refraction of light by cornea, aqueous humor, lens, and vitreous can be traced from Al Hazan to the modern focus in Von Helmholtz.[1] Insight into the transduction of light into neural impulse began last century with the studies of Young[2] and continues today, the quest now being nothing less than the understanding of vision itself. It has taken longer for the side effects of light to attract attention, but as this book testifies, this is now a very active field, with many publications on light-induced damage to cornea, lens, and retina. But away from the excitement of the optic axis, in the quiet and shaded recesses of the chamber angle, is light of any consequence? The purpose of this chapter is to explore this possibility.

In nature, there are two ways for light to reach the tissues of the chamber angle: scatter from sources in the media of the anterior segment and passage through the overlying limbal tissues. Normal sclera attenuates light about 40-fold at medium wavelengths[3] and probably more at short.[4] Although in the limbal area the effective scleral thickness is somewhat reduced, it is unlikely that it is reduced enough to allow significant amounts of light to pass through. The normal anterior segment is optically clear and scatters little light. Even in the presence of heavy "cells and flare," the amount of light scattered into the angle must be negligible compared with the intensities of direct illumination. Therefore, in seeking the effects of light on angle structures we are most likely to find indirect effects, most obviously, the effects of reactive products of light-catalyzed reactions in the aqueous humor.

Oxygen

Light may be directly damaging to various biological compounds, but the effect is small compared with the damage mediated by oxygen. This element is second only to fluorine in its electronegativity.[5] It is the quantum spin restriction that renders dioxygen stable in the biological environment. With enough energy, the kinetic barrier that this creates can be overcome, at Fahrenheit 451, or in the presence of photons.

Antoinette Pirie was among the first to draw attention to the presence of a reactive oxygen product in the aqueous humor.[6] She demonstrated that hydrogen peroxide was present in the aqueous humor of rabbit and cattle eyes and suggested that it was formed by the oxidation of ascorbate by O_2, catalyzed by light and traces of riboflavin. Recent studies by Varma and co-workers showed that reactive oxygen species were generated by light in cultured lenses, even in the absence of riboflavin.[5] Conversely, in the presence of micromolar concentrations of copper or iron complexes, ascorbate can reduce O_2 in the dark.[7] The first step in the reduction of oxygen is the formation of superoxide. This species is highly reactive in an aprotic environment, such as a cell membrane, but it is stabilized in a protic solvent such as water. In pure water the only pathway of removal is the self-dismutation:

$$O_2^{\cdot-} + HO_2^{\cdot} -> HO_2^{-} + O_2.^{8}$$

However, the aqueous humor of diurnal mammals, including man, has a high level of ascorbate,[9,10] which acts as a superoxide scavenger.[7,11] No estimates are available on the lifetime of superoxide in aqueous humor, but it would appear unlikely that it persists very long in the presence of millimolar levels of ascorbate.

There are other even more damaging species of reduced oxygen, in particular the hydroxyl radical, which is produced by the one-electron reduction of H_2O_2. A number of studies suggest that it is possible to generate this radical in model systems; for example: Rowley and Halliwell found that copper at physiological levels would catalyze the reduction of hydrogen peroxide by either superoxide or ascorbate, giving rise to $OH^{\cdot}$ or "species of equivalent reactivity."[12] Other studies by Halliwell and co-workers showed that hyaluronate could be depolymerized by ascorbate and O_2 or by ascorbate and H_2O_2 in the presence of Fe^{2+}. This reaction was blocked by catalase and hydroxyl radical scavengers, but not by superoxide dismutase.[13,14] However, lipid peroxidation in cultured rat lenses exposed to light could be blocked by the addition of superoxide dismutase, catalase, or ascorbate.[5,15] There is probably a superoxide-dependent and a superoxide-independent pathway of hydroxyl radical formation.[16] Most of the evidence for hydroxyl radical involvement in cellular damage is indirect, involving the use of scavengers whose specificities are always open to question. However, recently Varma and co-workers showed that in the cultured lens system O_2^- and H_2O_2 dependent damage was accompanied by the formation of $OH^\cdot$, as measured by electron spin resonance.[17] Nevertheless, all these model systems deviate substantially in one respect or another from physiological conditions, and it is still far from certain that there is significant involvement of $OH^\cdot$ in tissue damage in vivo.[16,18] In addition, the scavenging action of ascorbate on superoxide applies *a fortiori* to hydroxyl radical. Therefore it is very unlikely that any light-generated $OH^\cdot$ reaches the angle of the eye. In sum, the action of light, oxygen, and ascorbate in aqueous humor is to form superoxide, which is then converted by further reduction with ascorbate to hydrogen peroxide. The route to hydrogen peroxide may or may not be complicated by side reactions generating highly reactive species such as hydroxyl radical. The further reduction of H_2O_2 by ascorbate proceeds very slowly in aqueous humor,[10] consequently H_2O_2 accumulates.

Hydrogen peroxide has been measured in aqueous humor by a number of workers.[9,10,19,20] The level is remarkably high, about 25 μM in calf, man, and non-human primate; in guinea-pig it is about 47 μM and levels in rabbit have been variously measured from 33 to 85 μM. In contrast, the level in rat, a nocturnal animal, was, 9 μM. Giblin and co-workers found that levels of H_2O_2 in rabbit aqueous were highly correlated with ascorbate levels. An even more convincing correlation was shown by lowering the aqueous levels of ascorbate in guinea-pig tenfold by means of a scorbutic diet and finding a corresponding decrease in the level of H_2O_2.[9] This parallel behavior is consistent with ascorbate being the initiator of oxygen reduction in aqueous humor. The low levels of ascorbate and H_2O_2 in nocturnal mammals strongly support the role of light in catalyzing the reduction of oxygen. The pattern in birds, however, is puzzlingly different. The levels of ascorbate found by Reiss and co-workers in the aqueous humor of diurnal birds were no higher than those found in nocturnal birds.[21] The levels of H_2O_2 were not reported.

While the more reactive products of oxygen reduction are unlikely to reach the angle, if these species are formed, they may well attack constituents of the aqueous humor or lens, iris, or corneal membranes before being quenched by the ascorbate. Hydroxyl radical, for instance, can react with proteins, lipids, and nucleic acids in a variety of ways, being capable of attacking even saturated aliphatic side chains.[18] These reactions inside the cell are damaging; outside the cell they are scavenging as long as the reaction products themselves are not toxic. Most probably are not; but lipid peroxides, hydroperoxides, and malondialdehyde, all products of the oxidation of polyunsaturated fatty acids, possibly are. Malondialdehyde levels of aqueous humor have not been measured, but Cotlier and co-workers found lipid hydroperoxide levels of 45 to 105 μM in human, monkey, and rabbit,[22] suggesting that there is active oxidation of anterior chamber lipid. What then do the tissues of the chamber angle have to fear from the field of light? Hydrogen perioxide certainly, probably lipid hydroperoxide, and possibly malondialdehyde.

Besides photon activation there are other sources of reduced oxygen species, notably "leakage" from mitochondrial respiration, the "killing reaction" of phagocytic cells,[23] and reduction by melanin particles.[24] The principal respiratory leakage product is superoxide, and, not surprisingly, mitochondria contain large amounts of the protective enzymes, superoxide dismutase[25] and glutathione peroxidase.[26] There is little suggestion that these leakage substances cause significant cell damage. Phagocytosis by neutrophils, monocytes, macrophages, and endothelial cells is accompanied by the release of a variety of reduced oxygen species, $O_2^{\cdot}$, H_2O_2 and possibly $OH^{\cdot}$. A destructive compound, seemingly unique to the killing reaction, is hypochlorous acid, a product of the enzyme myeloperoxidase. The mechanism of the killing reaction is slowly being elucidated but it has yet to be fully understood.[27–29] There is growing evidence that inflammatory cells, in particular neutrophils, can seriously damage endothelium, both in tissue culture[30,31] and in intact tissue.[32] Such reactions are not "light-induced," but the toxic agents are mostly in common. Therefore, the study of inflammatory damage may well illuminate the mechanism of light-induced damage to the angle of the anterior chamber.

Angle Structures

The principal tissue of the angle is the trabecular meshwork (TM): the drain of the eye. However, it is not a gurgling drain. In the human eye, the linear rate of flow through the trabecular face is about 1.6 μm/s.[*] That is about five seconds per red cell diameter. At these low flow rates, diffusional transport becomes very important. The self-diffusion coefficient of water in water is about 2,800 μm^2/s.[33] The reaction products of oxygen are of similar molecular weight, and the diffusion coefficients will be the same to within a factor of two or so. Assuming the distance from the face of the TM to the edge of the limbal umbra to be approximately 500 μm, then the Péclet number is given by $\dfrac{500 \times 1.6}{2,800} = 0.3$. A Péclet number of 1 would indicate that diffusion and flow were equally important in transporting material to the TM face. A value very much less than 1 would indicate that diffusion completely dominated the process. The value of 0.3 that we obtain here indicates that transport is slightly dominated by diffusion. The lifetime of H_2O_2 in aqueous humor is sufficiently long[10] that diffusion alone should essentially equilibrate the concentration at the TM face with the bulk of aqueous humor.

At first sight the TM is a rather unexciting band, discernible in the gonio-lens only after considerable practice. Under the electron microscope, it appears to be little more than its name implies, a mesh of tissue functioning as a grating over the drain. But the TM has a disease, glaucoma, that in industrialized countries, is second only to accident in robbing people of their sight. Stimulated and funded by this medical interest, the biologist soon finds that this few-hundred-microns-high by a few-hundred-microns-deep thread of tissue holds many an unanswered riddle; to give only a few: What is the principal site of normal resistance to outflow?— of pathological resistance to outflow? Why when it is so difficult to damage the TM experimentally, do people still get glaucoma? What are the morphological changes that cause glaucoma (as distinct from those induced by treatment) and what biochemical changes cause the morphological changes?

In an area so beset by ignorance, we are inevitably subject to fashion. The new look in glaucoma pathogenesis is "oxidative damage." This idea was generated in part by findings within the field of TM biochemistry[34-38] and in part by findings in cataract research.[5,39-41] Each previous fashion has stimulated the production of much useful information, but none so far has answered the biochemical riddle. Naturally we who are working in the field of oxidative damage hope that we have found the key.

---

[*] Ross Ethier, Massachusetts Institute of Technology, unpublished calculations, 1986.

Sulfhydryls

Our story now requires a brief digression into some TM biochemistry. In 1953 Ernst Barany was investigating the question: Does aqueous humor leave the eye through the TM by an energy-requiring mechanism?[42] He perfused calf eyes with a variety of toxic agents, including cyanide, fluoride, arsenite, and iodoacetate. None had any effect on facility of aqueous outflow except iodoacetate, which caused a marked *increase* in outflow. In 1974 Van Buskirk and Grant[43] perfused eyes at various temperatures, the hypothesis being that by lowering the temperature, metabolic processes would be slowed and a pronounced effect on facility would be seen. However, no effect was detected beyond that accountable by the increase in viscosity of water. Since these two studies, it has been generally accepted that aqueous leaves the anterior chamber by bulk flow through the TM and not by any energy-requiring process. This left Barany's observation on iodoacetate as a puzzling and isolated observation. Starting in 1981, this laboratory has confirmed and extended this original finding. Calf eyes were perfused with a doses of 22 to 90 μmol of iodoacetamide. At the lowest dose there was no detectable change in the outflow facility. At 45 and 90 μmol there were substantial increases. However, when the TM from these eyes were dissected and homogenized, it was found that even at the lowest dose the capacity of the TM cells to carry out glycolysis was completely inhibited.[34] This suggested that the effect on outflow facility was not dependent on the metabolic effects of iodoacetamide. These effects on facility were also seen in baboon and in an assortment of non-baboon primate eyes. A year later, we published a similar series of perfusions, this time using N-ethylmaleimide as the -SH reactive agent.[35] Increases in facility of up to 20% were seen in calf eyes and even higher in some monkey eyes. In contrast to the effects with iodoacetamide, homogenates of TM from perfused calf eyes showed much lower levels of inhibition of glycolysis. Effects on facility could be clearly seen at doses that produced only 10% inhibition of lactate production from glucose. Given that glyceraldehyde 3-phosphate is among the most sensitive enzymes to N-ethylmaleimide, these results reinforced our suspicion that the effect on facility was not mediated by an effect on intracellular metabolism but rather by an effect on the cell membranes of the TM endothelium. A morphometric study of the TM in monkey eyes after perfusion with these two agents showed that even cells that must be considered dead, given the doses of agent to which they were exposed, showed little obvious morphological change.[36] However, by determined application of the digitizing pad to many electron microscope (EM) pictures, it was teased out that the principal effect of iodoacetamide was to loosen the attachment of the endothelial lining of Schlemm's canal to the underlying juxtacanilicular meshwork. There was no increase in the number of breaks in cell-to-cell junctions, nor any decrease in the number of "giant vacuoles," indicating that there was no decrease in the pressure drop across the inner wall of Schlemm's canal. In contrast, the principal effect of N-ethyl-maleimide was to cause just such cell breaks and a concomitant decrease in the number of "giant vacuoles." It would seem that quite subtle changes in morphology can produce measurable effects on the facility of outflow. Extending the observations on sulfhydryl agents to p-chloromercuribenzoate (PCMB) and its sulfonate p-chloromercuribenzene sulfonate (PCMBS) produced something of a surprise, although in retrospect, perhaps it should not have been.[37] Both monkey and calf eyes showed a marked *decrease* in outflow facility when perfused with these agents. Biochemical checks on excised TM from perfused eyes showed that in most cases glycolytic capacity was essentially unaffected. This is to be expected from agents that are often chosen because they do *not* enter cells readily. Morphological study of the primate eyes revealed extensive cell swelling, and it was this swelling that we felt to be responsible for the decrease in facility. Calculations by Ross Ethier of the Massachusetts Institute of Technology show that if an allowance is made for the known glycosaminoglycan content of TM, then the calculated and observed fractional decreases are very close.[44] The mechanism of the swelling probably involves an attack on one or more of the membrane adenosine triphosphatases (ATPase). Similar effects are seen in other cells such as red cell,[45] Ehrlich ascites tumor cells,[46] and renal tubular endothelium.[47] Meanwhile, what about more physiological agents that might have an effect on the membrane sulfhydryls of TM?

Hydrogen Peroxide

As noted above, the aqueous humor of diurnal mammals has high levels of ascorbate and H_2O_2, whereas that of dirunal birds does not. It is commonly supposed that the ascorbate is present to reduce active oxygen species created by irradiation with light. How then do the diurnal birds, whose need for such protection may be even greater, manage to survive the damaging effects of light? Why do mammals use ascorbate for this purpose when ascorbate is the principal generator of superoxide and hydrogen peroxide in aqueous humor? Perhaps the H_2O_2 is not an unfortunate by-product, but what nature selected for in the first place. This would make more sense of the high ascorbate/H_2O_2 levels in the rabbit whose crepuscular behavior would limit its exposure to strong light. Perhaps it emerges to catch the last rays of daylight to boost its H_2O_2 levels before nightfall. The principal argument against H_2O_2 being a desired product is the lack of any known function. The concentration, while high for a biological tissue, is not high enough to be disinfectant. If it is to regulate glycosamino-glycan levels by degrading them, there is the problem of how nocturnal mammals carry out this function. Regardless of whether the H_2O_2 is there accidentally or by "intent," it is still potentially damaging and our laboratory set out to investigate this question.

In a series of perfusions of calf eyes with various levels of hydro-gen peroxide, Michael Kahn showed that it had no detectable effect on the facility of outflow up to 25 mM, some 1,000-fold the normal concentration.[38] However, if he first inhibited gluta-thion reduction by inhibiting glutathione reductase with 1,3 *bis*(2-chloroethyl)-1-nitrosourea (BCNU) and then destroyed existing glutathione with diamide, then 25 mM hydrogen peroxide could damage the TM sufficiently to cause a decrease in outflow facility. Of course what these experiments did not account for was that TM cells may well have had a second line of defense, catalase. Subsequent work of ours, described below, showed that this in-deed was true. In 1984 Jon Polansky and co-workers reported that if catalase was inhibited with 3-aminotriazole, human TM cells in culture would show morphological evidence of damage after exposure to 100 μM H_2O_2.[48] In a series of recent experiments, Khiem Nguyen repeated the perfusion experiments with the addi-tion of 20 mM aminotriazole to inhibit catalase. One did not need EM and a digitizing pad to detect the damage at 25 mM hydrogen peroxide. Almost immediately the tissues of the anterior chamber began to distintegrate; the iris fell apart; the anterior chamber blackened with released pigment—at once a demonstration of the impressive destructive powers of hydrogen peroxide and of the impressive preservative powers of catalase.

Defensive Enzymes

Spurred on by the findings of Michael Kahn, we embarked on a series of studies of the protective enzymes of the TM. These were carried out by a succession of energetic students and fellows. The first was Doug Scott,[49] who demonstrated the presence of gluta-thione peroxidase (GPx) at a level of 0.60 μmole/min/g wet wt. The K_m for H_2O_2 was found to be 12 μM, about half the concentra-tion in aqueous, but the K_m for glutathione, reduced form (GSH) was 2.9 mM with pronounced sigmoidal kinetics. This K_m is high compared with the tissue content of GSH of 0.4 μmole/g wet wt.[38] Even after allowing for the fact that only half the wet weight is cellular,[†] and therefore the true concentration is probably near 1 mM, this is still a rather higher K_m than one might expect. Perhaps in its intracellular environment the curve is shifted leftward by allosteric modifiers or protein-protein interaction. The sigmoidicity serves to prevent a sudden influx of H_2O_2 stripping the cell of all its GSH. A prudent provision for a resource required by many other reactions.

The second worker was Howard Weiss, who tackled glutathione reductase (GR).[50,51] He demonstrated a level of 0.12 μmole/min/g wet wt. The study of this enzyme was complicated by rapid loss of activity. Howard succeeded in producing a stable preparation by heat treatment, which presumably destroyed an inactivating protease. Khiem Nguyen took up the project and purified the enzyme further by affinity chromatography and determined K_ms of 19 μM for NADPH and 78 μM for GSSG. The activity of GR is only one fifth that of GPx. However, this apparent mismatch is in maximal activities. Actual fluxes are governed by demand. There is a large reservoir of GSH that may be drawn down by different reactions at different times, while GR works constantly at replenishment until the reservoir is restored.

Khiem Nguyen has also examined glucose 6-phosphate dehydrogenase.[52] The tissue activity was found to be 0.23 μmole/min/g wet wt. Activity stain on gel electrophoresis revealed a complex pattern of isoenzymes with six to seven bands of varying intensities. Because of this mixture of types, no effort was made to determine kinetic parameters. However, the physiological response to the NADPH/NADP$^+$ ratio was determined. The activity declined more or less linearly from 100% of maximal at a ratio of 0 to 0% at a ratio of 10. The ratio in vivo is usually about 5,[53] which is in the midpoint of the regulatory curve. The activity of the second enzyme of the hexose monophosphate pathway, 6-phosphogluconate dehydrogenase, was also measured and found to be 0.47 μmole/min/g wet wt.[53a]

The catalase and superoxide dismutase of calf TM were studied by Sharon Freedman.[54] The catalase activity was 0.88 μmole/min/g wet wt, which is comparable to the levels found in iris and retina and much higher than that found in lens. This remarkable enzyme is the fastest known, its rate of reaction being essentially diffusion limited.[55] This means that it has no definable K_m. Activity stain on gel electrophoresis revealed only one band. Superoxide

---

[†] Andrew Dorfman, unpublished data, 1984.

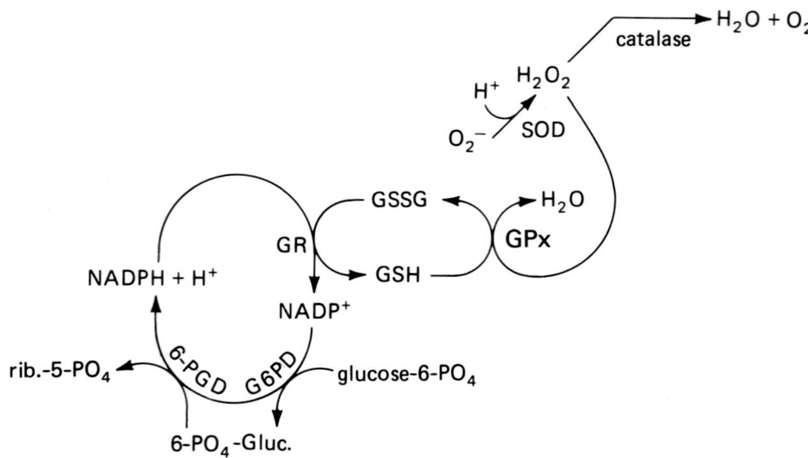

Figure 3.1. Interrelationship of the enzymes defensive against oxidative damage. SOD: superoxide dismutase; GPx: glutathione peroxidase; GR: glutathione reductase; G6PD: glucose 6-phosphate dehydrogenase; 6-PGD: 6-phosphogluconate dehydrogenase. All these enzymes have been found and measured in calf trabecular meshwork.

dismutase content was found to be 0.18 "unit"/mg wet wt. (The method of measurement of this enzyme yields no natural unit such as μmole/min. This means that every group defines its own unit, which cannot always be related to those of others.) Activity staining on gel electrophoresis showed three isoenzymes, one of which was insensitive to cyanide and was probably the manganese-containing mitochondrial enzyme. The other two were cyanide-sensitive and were probably the copper-zinc cytosolic enzyme. Superoxide dismutase activity has also been identified in the TM

Table 3.1. Enzymes of TM defensive against oxidative damage.

Enzyme	Tissue			
	TM	Retina	Liver	Reference
Catalase [*]	0.88	0.45	63	54
Superoxide dismutase [*]	0.18	0.17	1.3	54
Glutathione peroxidase [†]	0.60	—	—	49
Glutathione reductase [†]	0.12			51
Glutathione reductase [‡]	5	—	120	
Glucose 6-phosphate dehydrogenase [†]	0.23			53A
Glucose 6-phosphate dehydrogenase [‡]	10	5	5	
6-Phosphogluconate dehydrogenase [†]	0.47	—	—	53A

[*] "Units"/mg wet wt

[†] μmol/min/g wet wt

[‡] nmol/min/mg soluble protein

from adult cattle, adult and neonatal cats, and rabbits.[‡] The interrelationships of these enzymes are summarized in Figure 3.1.

This then is the battery of enzymes defensive against oxidative damage (Table 3.1). We have seen already how impressive the excess capacity of this battery is in defending against H_2O_2, at least in neonatal calf.

Glaucoma

Can oxidative damage cause glaucoma? At first glance, it would seem unlikely unless something goes very wrong. A possible example of something going very wrong is glucose 6-phosphate dehydrogenase deficiency, a condition found mostly in blacks and in some Mediterranean races. There is a study linking this condition with an excess incidence of cataracts.[56] No association has been found with glaucoma, but this may simply be because none has been looked for. A small case-control study was carried out at the Massachusetts Eye and Ear Infirmary to determine if such an association could be detected. The results were inconclusive. However, it would be very valuable to have other centers take up this question also.

Nature has not been so obliging in providing the investigator with deficiencies of catalase, superoxide dismutase, GPx, or GR to study. But more importantly, the majority of sufferers from glaucoma do not have in common a known enzyme deficiency. If H_2O_2 is the (or a) cause of primary open-angle glaucoma, how can it damage the TM to the point of serious dysfunction when the TM appears to be so well defended? At this point, one can only speculate about possible mechanisms.

The TM shows its age by a steady depletion of endothelial cells.[57] However, most people do not get glaucoma. Presumably those who suffer from this disease are exposed to a higher level of insult or have a decreased capacity for defense compared with those who are resistant. Cataract provides us with an example of a disease caused by decreased capacity for defense in the glucose 6-phosphate dehydrogenase deficiency alluded to above, and with an example of the role of increased insult: there appears to be a correlation between light exposure and senile cataract.[39] As yet, no such associations have been established for glaucoma. At this stage we must content ourselves with an examination of possible mechanisms of damage and leave aside the question of insult versus susceptibility.

[‡] Richard Bensinger, personal communication, 1985.

Hydrogen peroxide passes over the outside of TM endothelial cells whereas catalase and GPx are on the inside. Membrane constituents and/or components of the extracellular matrix that are damaged by H_2O_2 presumably are internalized and repaired or recycled. For instance, GPx can reduce not only H_2O_2 but many organic peroxides as well.[26] Perhaps with age the cell's biosynthetic capacity cannot keep up with the constant rapid turnover required by this hostile environment. Or perhaps there is a breakdown in this process, say by accumulation of small amounts of nondegradable debris; for instance, the cumulative effect of lipid peroxidation is a perennial favorite as a mechanism of aging. Leibowitz and Siegel claim that ascorbate levels decline with age in humans, guinea-pig, and mouse and that synthetic antioxidants prolong life span in the experimental animals.[58] While all animals age, unfortunately none but man get spontaneous primary open-angle glaucoma. In addition, it is known that H_2O_2 reacts poorly with membrane lipid.[59] A more likely target is the membrane proteins. Garner and Spector have shown that in cataract formation the membrane proteins are oxidized before other cell proteins,[60] and the same authors have suggested that the primary damage in cataract formation may be the uncoupling of the ATPase-driven pumps in the lens epithelial cells.[61,62] This suggestion prompted an investigation by us of TM plasma membrane ATPases. Khiem Nguyen isolated purified plasma membrane vesicles from calf TM and characterized the principal ATPase.[§] The *ATPase* activity of Na^+/K^+ ATPase is susceptible to inhibition by PCMB, PCMBS, and ouabain, but H_2O_2 is largely without effect. The effect of H_2O_2 on *pumping* activity has yet to be studied. Some other interesting effects have been noted. Jon Polansky and co-workers found that the normal release of PGE_2 by cultured human TM cells is inhibited by exposure of the cells to physiological levels of H_2O_2 ($3 - 6 \times 10^{-5}$ M) if catalase was first inhibited by 3-aminotriazole. At 3×10^{-4} M H_2O_2 the same response could be elicited even in the absence of 3-aminotriazole.[‖] Similar effects were noted by Ager and Gordon, who found H_2O_2 could stimulate or inhibit PGE_2 release by cultured pig aortic endothelium, depending on the conditions used.[63] In sum, it has not been possible so far to show acute damage to TM in any of the systems studied except at very high levels of H_2O_2. If the defensive systems are inhibited, damage can be shown at somewhat lower levels, and some physiological responses can be demonstrated at physiological levels.

[§] Khiem Nguyen, manuscript in preparation, 1986.
[‖] Jon Polansky, personal communication, 1985.

Glaucoma is, however, a chronic disease, not an acute one. The failure to show effects of H_2O_2 on TM in the short term does not weigh too heavily against H_2O_2 as a possible etiological agent. The effects could be slow and subtle. E.I. Anderson and co-workers showed some years ago that the corneal endothelium requires not only GSH but GSSG for proper functioning of endothelial pumps.[64] The dramatic effects of sulfhydryl-active reagents on the facility of aqueous outflow suggest that perhaps a similar "sulfhydryl tone" may be required for the proper functioning of the TM. A small displacement over the years could conceivably result in an eventual breakdown in some essential cellular capacity or property. This breakdown need not even be enzymatic, let alone be specifically of an ATPase. It could be an effect on the properties of the cell membrane, such as the one recently described by Tripathi et al, where endothelial cells from normal and glaucomatous eyes differed in their distribution of sialated glycoproteins.[65]

Another source of active oxygen species is inflammatory cells. There is considerable evidence that leakage of substances of the "killing reaction" may damage adjacent tissues.[31,66,67] Primary open-angle glaucoma does not appear to be an inflammatory disease, nor are the products of the "killing reaction" light-induced; but inflammation, as in chronic uveitis, can cause damage to the TM, with a resulting secondary glaucoma.[68] Perhaps this could be considered a speeded up oxidative damage model of primary open-angle glaucoma. In addition, active oxygen species such as $O_2^{\cdot-}$ generated intravitreally can induce or augment an inflammatory reaction.[69] Superoxide is known to generate factors chemotactic for neutrophils. (Perhaps the leukotrienes were evolved from the battle-smoke of phagocytosis.)

Phagocytic cells can not only injure adjacent cells but may destroy themselves.[70] Trabecular meshwork endothelium is phagocytic.[71–73] In its role of the "self-cleaning filter"[74] it ingests various kinds of debris, including pigment. Melanin itself is capable of generating active oxygen species in the dark and at a much greater rate in the presence of light.[24] Perhaps all primary open-angle glaucoma is pigmentary. With normal loads of pigment, only the few individuals susceptible get the disease; with increased load, such as in pigmentary dispersion syndrome, the fraction susceptible increases and the onset is earlier. Arguing against this idea is the finding of Zink and co-workers that patients with pigmentary glaucoma lacked the heightened steroid sensitivity found in primary open-angle glaucoma.[75]

It is readily seen that oxidative damage as a mechanism of primary open-angle glaucoma is long on speculation and short on data. Apart from some data on the protective enzymes and a few in vitro observations of H_2O_2 damage to TM or TM cells in culture, there is precious little information.

What Is to be Done?

The following are some suggestions:

1. For more glaucoma centers to carry out a search for an association between glucose 6-phosphate dehydrogenase deficiency and glaucoma.
2. To measure, in vivo, and/or in model systems, the generation and decay of oxygen metabolites in aqueous humor.
3. To measure the capacity of TM to remove H_2O_2 from aqueous humor and by selective inhibition to determine the enzymes responsible.
4. Develop measures of oxidative damage, such as changes in rubidium uptake by TM cells, using perfused eyes, isolated cell suspensions, and tissue culture.
5. To further identify and characterize the defensive enzymes in animal TM and in human, using tissue culture and where possible excised tissued.
6. The most pressing need, and the most difficult, is to develop an in vivo model of primary open-angle glaucoma.

But what is the definitive experiment? The Michelson-Morley experiment of the oxidative damage hypothesis? It is most unlikely that there will ever be one. Glaucoma is a chronic disease, and our experiments are acute. Like the cancer researcher feeding the equivalent of 800 cans of diet soda per day, we too must speed up the pathogenetic process by hobbling the defensive enzymes, or by raising the level of insult. Our methodology therefore is inherently flawed and our interpretations are always open to objection. Plausibility is the most that we can hope for. However, mere plausibility isn't so bad if we can show that real human TM from patients with glaucoma mimics our pathogenetic models. Demonstrating this will be extremely difficult. To have enough material, it will require a biochemistry of trabeculectomy specimens. It may be possible with present techniques to measure semiquantitatively enzyme levels and even isoenzyme types by immunocytochemistry, using monoclonal antibodies. Metabolic studies on undissected specimens would require something like imaging nuclear magnetic resonance, but the needed resolution and sensitivity are far beyond present capability. The alternative of microdissection and enzymatic cycling assays [76] is arduous and time consuming and should perhaps be reserved for answering the few critical questions when these have been established by other methods.

Conclusion

One day it may be demonstrated to the general satisfaction that glaucoma is caused by a failure to cope with the chronic insult of H_2O_2. If this comes to pass, then glaucoma must be added to the price the eye pays for being the organ of light.

References

1. Duke-Elder S, Abrams D: Ophthalmic optics and refraction, *in* System of Ophthalmology, vol 5, Duke-Elder S (ed). CV Mosby Co, St. Louis, pp 3–23, 1970.
2. Duke-Elder S, Gloster J: Physiology of the eye *in* System of ophthalmology, vol 4, Duke-Elder (ed). CV Mosby Co, St. Louis, pp 435–446, 1968.
3. Kopeiko LG, Koretskaya YM, Mitkokh DI, Chentsova OB: Spectral characteristics of the eyeball coat. Vestn Oftalmol 1:46–49, 1979.
4. Spillman L: Density, light scatter, and spectral transmission of a scarred human cornea. Albrecht Von Graefes Arch Klin Exp Ophthalmol 184:278–286, 1972.
5. Varma SD, Chand D, Sharma YR, Kuck JF Jr, Richards RD: Oxidative stress on lens and cataract formation: role of light and oxygen. Curr Eye Res 3:35–57, 1984.
6. Pirie A: Glutathione peroxidase in lens and a source of hydrogen peroxide in aqueous humour. Biochem J 96:244–253, 1965.
7. Scarpa M, Stevanato R, Viglino P, Rigo A: Superoxide ion as active intermediate in the autoxidation of ascorbate by molecular oxygen. Effect of superoxide dismutase. J Biol Chem 258:6695–6697, 1983.
8. Hill HAO: The chemistry of dioxygen and its reduction products, *in* Oxygen Free radicals and Tissue Damage, Fitzsimons DW (ed). Excerpta Medica, Amsterdam, pp 5–17, 1979.
9. Giblin FJ, McCready JP, Kodama T, Reddy VN: A direct correlation between the levels of ascorbic acid and H_2O_2 in aqueous humor. Exp Eye Res 38:87–93, 1984.
10. Spector A, Garner WH: Hydrogen peroxide and human cataract. Exp Eye Res 33:673–681, 1981.
11. Som S, Raha C, Chatterjee IB: Ascorbic acid: a scavenger of superoxide radical. Acta Vitaminol Enzymol 5:243–250, 1983.
12. Rowley DA, Halliwell B: Superoxide-dependent and ascorbate-dependent formation of hydroxyl radicals in the presence of copper salts: a physiologically significant reaction? Arch Biochem Biophys 225:279–284, 1983.
13. Rowley DA, Halliwell B: Formation of hydroxyl radicals from hydrogen peroxide and iron salts by superoxide- and ascorbate-dependent

mechanisms: relevance to the pathology of rheumatoid disease. Clin Sci 64:649–653, 1983.

14. Wong SF, Halliwell B, Richmond R, Skowroneck WR: The role of superoxide and hydroxyl radicals in the degradation of hyaluronic acid induced by metal ions and by ascorbic acid. J Inorg Biochem 14:127–134, 1981.

15. Varma SD, Srivastava VK, Richards RD: Photoperoxidation in lens and cataract formation: preventive role of superoxide dismutase, catalase and vitamin C. Ophthalmic Res 14:167–175, 1982.

16. Winterbourn CC: Hydroxyl radical production in body fluids. Roles of metal ions, ascorbate and superoxide. Biochem J 198:125–131, 1981.

17. Varma SD, Richards RD, Bolton T, Rice D: Mechanism of hydrogen peroxide damage to the lens in vitro. Invest Ophthalmol Vis Sci 26[Suppl]:295, 1985.

18. Willson RL: Hydroxyl radicals and biological damage in vitro: what relevance in vivo? *in* Oxygen Free Radicals and Tissue Damage, Fitzsimons DW (ed). Excerpta Medica, Amsterdam, pp 19–42, 1979.

19. Matsuda H, Giblin FJ, Reddy VN: The effect of x-irradiation on cation transport in rabbit lens. Exp Eye Res 33:253–265, 1981.

20. Bhuyan KC, Bhuyan DK: Regulation of hydrogen peroxide in eye humors. Effect of 3-amino-1H-1,2,4-triazole on catalase and glutathione peroxidase of rabbit eye. Biochim Biophys Acta 497:641–651, 1977.

21. Reiss GR, Werness PG, Brubaker RF: Aqueous ascorbic acid levels in diurnal birds. Invest Ophthalmol Vis Sci 26[Suppl]:101, 1985.

22. Cotlier E, Panahbarhagh H, Obara Y: Lipid hydroperoxide formation by human aqueous humor, by cataracts, and in diabetic rabbits. Inv Ophthalmol Vis Sci 26[Suppl]:295, 1985.

23. Fridovich I: Chairman's introduction, *in* Oxygen Free Radicals and Tissue Damage, Fitzsimons DW (ed). Excerpta Medica, Amsterdam, pp 1–4, 1979.

24. Sarna T, Duleba A, Korytowski W, Swartz H: Interaction of melanin with oxygen. Arch Biochem Biophys 200:140–148, 1980.

25. Fridovich I: Superoxide dismutases: defence against endogenous superoxide radical, *in* Oxygen Free Radicals and Tissue Damage, Fitzsimons DW (ed). Excerpta Medica, Amsterdam, pp 77–93, 1979.

26. Flohe: Glutathione peroxidase: fact and fiction, *in* Oxygen Free Radicals and Tissue Damage, Fitzsimons DW (ed). Excerpta Medica, Amsterdam, pp 95–122, 1979.

27. Segal AW, Allison AC: Oxygen consumption by stimulated human neutrophils, *in* Oxygen Free Radicals and Tissue Damage, Fitzsimons DW (ed). Excerpta Medica, Amsterdam, pp 205–223, 1979.

28. Roos D, Weening RS: Defects in the oxidative killing of microorganisms by phagocytic leukocytes, *in* Oxygen Free Radicals and Tissue Damage, Fitzsimons DW (ed). Excerpta Medica, Amsterdam, pp 225–262, 1979.

29. Rosen GM, Freeman BA: Detection of superoxide generated by endothelial cells. Proc Natl Acad Sci USA 81:7269–7273, 1984.

30. Sacks T, Moldow CF, Craddock PR, Bowers TK, Jacob HS: Oxygen radicals mediate endothelial cell damage by complement-stimulated granulocytes. An in vitro model of immune vascular damage. J Clin Invest 61:1161–1167, 1978.

31. Weiss SJ, LoBuglio AF: Phagocyte-generated oxygen metabolites and cellular injury. Lab Invest 47:5–18, 1982.

32. Steinberg H, Greenwald RA, Sciubba J, Das DK: The effect of oxygen-derived free radicals on pulmonary endothelial cell function in the isolated perfused rat lung. Exp Lung Res 3:163–173, 1982.

33. Refojo MJ: Permeation of water through some hydrogels. J Appl Polymer Sci 9:3817–3426, 1965.

34. Epstein DL, Hashimoto JM, Anderson PJ, Grant WM: Effect of iodoacetamide perfusion on outflow facility and metabolism of the trabecular meshwork. Invest Ophthalmol Vis Sci 20:625–631, 1981.

35. Epstein DL, Patterson MM, Rivers SC, Anderson PJ: N-ethylmaleimide increases the facility of aqueous outflow of excised monkey eyes. Invest Ophthalmol Vis Sci 22:752–756, 1982.

36. Lindenmayer JM, Kahn MG, Hertzmark E, Epstein DL: Morphology and function of the aqueous outflow system in monkey eyes perfused with sulfhydryl reagents. Invest Ophthalmol Vis Sci 24:710–717, 1983.

37. Freddo TF, Patterson MM, Scott DR, Epstein DL: Influence of mercurial sulfhydryl agents on aqueous outflow pathways in enucleated eyes. Invest Ophthalmol Vis Sci 25:278–285, 1984.

38. Kahn MG, Giblin FG, Epstein DL: Glutathione in calf trabecular meshwork and its relation to aqueous humor outflow facility. Invest Ophthalmol Vis Sci 24:1283–1287, 1983.

39. Chylack LT Jr: Mechanisms of senile cataract formation. Ophthalmology 91:596–602, 1984.

40. Megaw JM: Glutathione and ocular photobiology. Curr Eye Res 3:83–87, 1984.

41. Bhuyan KC, Bhuyan DK: Molecular mechanism of cataractogenesis: III. Toxic metabolites of oxygen as initiators of lipid peroxidation and cataract. Curr Eye Res 3:67–81, 1984.

42. Barany EH: In vitro studies of the resistance to flow through the angle of the anterior chamber. Acta Soc Med Upsaliensis 59:260–276, 1953.

43. Van Buskirk EM, Grant WM: Influence of temperature and the question of involvement of cellular metabolism in aqueous outflow. Am J Ophthalmol 77:565–572, 1974.

44. Kamm RD, Ethier CR, Freddo TF, Johnson MC, Epstein DL: The influence of changes in juxtacanalicular meshwork morphology on aqueous outflow resistance. Invest Ophthalmol Vis Sci 26[Suppl]:5, 1985.

45. Jacob HS, and Jandl JH: Effects of sulfhydryl inhibition on red blood cells: I. Mechanism of hemolysis. J Clin Invest 41:779–792, 1962.

46. Penttila A, Trump BF: Studies on the modification of the cellular response to injury: III. Electron microscopic studies on the protective effect of acidosis on p-chloromecuribenzene sulfonic acid-(PCMBS)

induced injury of Ehrlich ascites tumor cells. Virchows Arch B Cell Path 18:17–34, 1975.

47. Sahaphong S, Trump BF: Studies of cellular injury in isolated kidney tubes of the flounder: V. Effects of inhibiting sulfhydryl groups of plasma membrane with the organic mercurials PCMB (parachloromecuribenzoate) and PCMBS (parachloromecuribenzene sulfonate). Am J Pathol 63:277–298, 1971.

48. Polansky JR, Wood I, Maglio M, Addison J, Alvarado JA, Bhuyan KC, Bhuyan DK, Podos SM: Peroxide damage to human trabecular cells: a possible model for morphologic alterations in aging and glaucoma. Invest Ophthalmol Vis Sci 25[Suppl]:122, 1984.

49. Scott DR, Karageuzian LN, Anderson PJ, Epstein DL: Glutathione peroxidase of calf trabecular meshwork. Invest Ophthalmol Vis Sci 25:599–602, 1984.

50. Weiss HS, Karageuzian LN, Anderson PJ, Epstein DL: Glutathione reductase of calf trabecular meshwork. Invest Ophthalmol Vis Sci 25[Suppl]:206, 1984.

51. Nguyen KPV, Weiss H, Karageuzian LN, Anderson PJ, Epstein DL: Glutathione reducatase of calf trabecular meshwork. Invest Ophthalmol Vis Sci 26:887–890, 1985.

52. Anderson PJ, Nguyen KPV, Lee DA, Epstein DL: Glucose 6-phosphate DH of calf trabecular meshwork. Invest Ophthalmol Vis Sci 26[Suppl]:229, 1985.

53. Williamson DH, Brosnan JT: Concentration of metabolites in animal tissues, *in* Methods of Enzymatic Analysis, vol 4, Bergmeyer HU (ed). Academic Press, New York, pp 2266–2302, 1974.

53a. Nguyen K, Lee DA, Anderson PJ, Epstein DL: Glucose 6-phosphate dehydrogenase of calf trabecular meshwork. Invest Ophthalmol Vis Sci 27:992–997, 1986.

54. Freedman SF, Anderson PJ, Epstein DL: Superoxide dismutase and catalase of calf trabecular meshwork. Invest Ophthalmol Vis Sci 26:1330–1335, 1985.

55. Chance B, Sies H, Boveris A: Hydroperoxide metabolism in mammalian organs. Physiol Rev 59:527–603, 1979.

56. Harley JD, Robin H, Menser MA, Hertzberg R: Cataracts in G6PD deficiency. Br Med J 1:421, 1966.

57. Alvarado J, Murphy C, Polansky J, Juster R: Age-related changes in trabecular meshwork cellularity. Invest Ophthalmol Vis Sci 21:714–727, 1981.

58. Leibowitz BE, Siegel BV: Aspects of free radical reactions in biological systems: aging. J Gerontol 35:45–56, 1980.

59. Zimmermann R, Flohe L, Weser U, and Hartmann H-J: Inhibition of lipid peroxidation in isolated inner membrane of rat liver mitochondria by superoxide dismutase. FEBS Lett 29:117–120, 1973.

60. Garner MH, Spector A: Selective oxidation of cysteine and methionine in normal and senile cataractous lenses. Proc Natl Acad Sci USA 77:1274–1277, 1980.

61. Garner MH, Garner WH, Spector A: Kinetic cooperativity change

after H_2O_2 modification of (Na,K)-ATPase. J Biol Chem 259:7712–7718, 1984.

62. Garner WH, Garner MH, Spector A: H_2O_2-induced uncoupling of bovine lens Na^+, K^+-ATPase. Proc Natl Acad Sci USA 80:2044–2088, 1983.

63. Ager A, Gordon JL: Differential effects of hydrogen peroxide on indices of endothelial cell function. J Exp Med 159:592–603, 1984.

64. Anderson EI, Wright DD: Effects of S-methyl glutathione, S-methyl cysteine, and the concentration of oxidized glutathione on transendothelial fluid transport. Invest Ophthalmol Vis Sci 19:684–686, 1980.

65. Tripathi RC, Tripathi BJ, Spaeth G: Role of sialated glycoproteins in the aqueous outflow pathway. Invest Ophthalmol Vis Sci 26[Suppl]:110, 1985.

66. Perkowski SZ, Havill AM, Flynn JT, Gee MH: Role of intrapulmonary release of eiconsanoids and superoxide anion as mediators of pulmonary dysfunction and endothelial injury in sheep with intermittent complement activation. Cir Res 53:574–583, 1983.

67. Mittag T: Role of oxygen radicals in ocular inflammation and cellular damage. Exp Eye Res 39:759–769, 1984.

68. Grant WM: Glaucoma due to intraocular inflammation, *in* Glaucoma, Chandler PA, Grant WM (eds). Lea and Febiger, Philadelphia, pp 236–257, 1979.

69. Sery TW, Petrillo R: Superoxide anion radical as an indirect mediator in ocular inflammatory disease. Curr Eye Res 3:243–352, 1984.

70. McCord JM, Wong K: Phagocyte produced free radicals: roles in cytotoxicity and inflammation, *in* Oxygen Free Radicals and Tissue Damage, Fitzsimons DW (ed). Excerpta Medica, Amsterdam, pp 343–360, 1979.

71. Polansky JR, Wood IS, Maglio MT, Alvarado JA: Trabecular meshwork cell culture in glaucoma research: Evaluation of biological activity and structural properties of human trabecular cells in vitro. Ophthalmology 91:580–595, 1984.

72. Sherwood M, Richardson TM: Evidence for in vivo phagocytosis by trabecular endothelial cells. Inv Ophthalmol Vis Sci 19[Suppl]:66, 1980.

73. Sherwood M, Richardson TM: Kinetics of the phagocytic process in the trabecular meshwork of cats and monkeys. Inv Ophthalmol Vis Sci 20[Suppl]:65, 1981.

74. Bill A: The drainage of aqueous humor. Invest Ophthalmol Vis Sci 14:1–3, 1975.

75. Zink HA, Palmberg PF, Sugar A, Sugar HS, Cantrill HL, Becker B, Bigger JF: Comparison of in vitro corticosteroid response in pigmentary glaucoma and primary open angle glaucoma. Am J Ophthalmol 80:478–484, 1975.

76. Lowry OH, Passonneau JV: A flexible system for enzymatic analysis. Academic Press, New York, 1974.

4

Light and the Cornea and Conjunctiva

David Miller

The obvious resistance of the cornea to natural ultraviolet (UV)-induced damage seems to stand in direct contradiction to the laboratory data. (Parenthetically, one is reminded of a similar situation in which, according to aerodynamic principles, the honeybee should not be able to fly.) Results from a number of animal studies [1–5] can be used to predict that steady exposure to a combination of sun, sky, and reflective natural surfaces in a temperate climate should cause photokeratitis in 10 minutes. Obviously, this does not happen.

As yet, we have no proven solution to this dilemma. However, since most of these studies involved animals whose eyes were held open for prolonged periods of time, one might explore the unnatural aspects of such an experiment. Certainly, during that period the precorneal tear film is lost. It is possible that the normal tear film might hold back harmful UV radiation. Such speculation will have to await extensive in vivo studies. Thus, we are left with the knowledge that the cornea is resistant to damage caused by average levels of UV radiation.

Of course, the cornea can be damaged by excessive shortwave UV light, such as that produced by man-made sunlamps or arc-welding beams. And, yes, there are extreme natural situations such as the highly reflective and totally enveloping nature of arctic snow and ice, which causes actinic keratitis or Labrador keratitis in man. Then, too, it has been suggested that pterygium in man is the result of excessive and prolonged UV exposure. These entities are discussed in some detail, and a few unorthodox ways of looking at the topic are also given. In all, however, because of the efficient design of the cornea and conjunctiva, this chapter will be short.

Ultraviolet-Induced Keratitis

The subject of UV damage to the cornea was first studied in an objective and quantitative manner in 1916 by Voerhoeff et al.[1] They were the first to point out that the phenomenon was both wavelength and intensity dependent. Although their work involved rabbit corneas, subsequent work on primate corneas showed very similar findings.[3]

Since that time, the area of wavelength dependence has been thoroughly explored. The most effective range of wavelengths that produce corneal damage is between 260 and 290 nm.[2–4] Of course the sources of these wavelengths must be man-made machines, since the ozone layer allows essentially no UV light below 290 to enter our atmosphere. Nevertheless, let us focus on the quantitative relationship between wavelength and threshold energy needed to produce a corneal lesion. At about 270 nm, only 0.005 J/cm^2 of energy will produce a lesion. At 300 nm, about 0.01 J/cm^2 will produce a corneal lesion, or almost double the threshold energy. At 320 nm, 10 J/cm^2 produce a lesion, or an increase of energy of about 2,000 times above the lowest threshold. At 350 nm, the threshold dose is 50 J/cm^2; at 370 nm, about 100 J/cm^2; at 380 nm, about 150 J/cm^2; and at 390 nm, about 250 J/cm^2.[2–4] Two major groups of facts emerge from these studies. First, an incredibly small amount of short wavelength UV light can cause corneal damage. The sources of this form of UV light are man-made on earth, although astronauts above the ozone layer would also be vulnerable to short wavelength UV light. Biochemically speaking, the nucleic acids found in the epithelium and endothelium absorb maximally at these wavelengths as do certain aromatic amino acids such as tryptophan.[6] Since bacteria and virus also contain these compounds, they are very vulnerable to the killing effects of germicidal UV lamps. Second, natural solar UV-A (i.e., 320–400 nm) can produce corneal lesions, but the exposures clearly must be quite prolonged. Thus, snow blindness, i.e., actinic keratitis, would occur after prolonged UV exposure. Oddly enough, polar animals and birds do not appear to develop this condition.[8]

A few other interesting pieces of information concerning actinic keratitis have emerged from the literature. Long UV wavelength keratitis is oxygen dependent, and about twice as much energy is needed to produce a corneal lesion in a low-oxygen environment than in a high-oxygen environment.[9] In clinical terms one might predict that tight-fitting goggles or contact lenses make a patient less vulnerable to the development of UV-induced keratitis due to natural sunlight.

As suggested earlier, the primary site of damage in actinic keratitis is the epithelial layers and, to a lesser degree, the endothelium.[7,10] The histologic changes are not immediate and are most noticeable 12 hours after injury. These include epithelial cell swelling, death, and desquamation associated with keratocyte changes, stromal swelling, endothelial damage, and an anterior uveitis. Resolution takes two to seven days, depending on the severity of the burn. When the damaged epithelium sloughs away, the characteristic punctuate fluorescein staining seen with the slit lamp occurs. The endothelium, on the other hand, returns to health using a form of DNA repair.[11,12] It would appear that when the epithelial cells have sloughed away, one could say that a UV absorber is gone, and further exposure to UV light would allow much higher doses to strike the endothelial layer. Thus, it would seem reasonable to keep patients with large epithelial defects in sunglasses and/or away from sunny environments. Interestingly, in most cases of severe actinic keratitis, extreme photophobia and lid swelling appear to be nature's way of reducing further light damage during healing.

Labrador Keratopathy

Chronic exposure to the UV light of the Arctic environment appears to produce a condition known as Labrador keratopathy or spheroidal degeneration of the cornea. Careful epidemiologic research shows that the highest incidence occurs between 55 and 56 degrees north latitude, where the levels of reflected UV light are very high.[13] The incidence of this condition is about 14% in the Eskimo population.[14] This condition, first described clinically in 1965,[15] presents as a cornea-conjunctival lesion, which slowly creeps across the cornea in the intrapalpebral strip. The conjunctival lesion has engorged blood vessels, whereas the cornea develops a haze made up of droplet degeneration in the layer just beneath the epithelium extending from a clear limbal strip to the corneal center[16] (Fig. 4.1).

Pterygium

A typical pterygium is pictured in Figure 4.2. Chronic exposure to UV light has also been implicated in the causation of pterygium.[17–19] Evidence for this thesis is primarily epidemiologic, since areas of high pterygium incidence and those receiving high

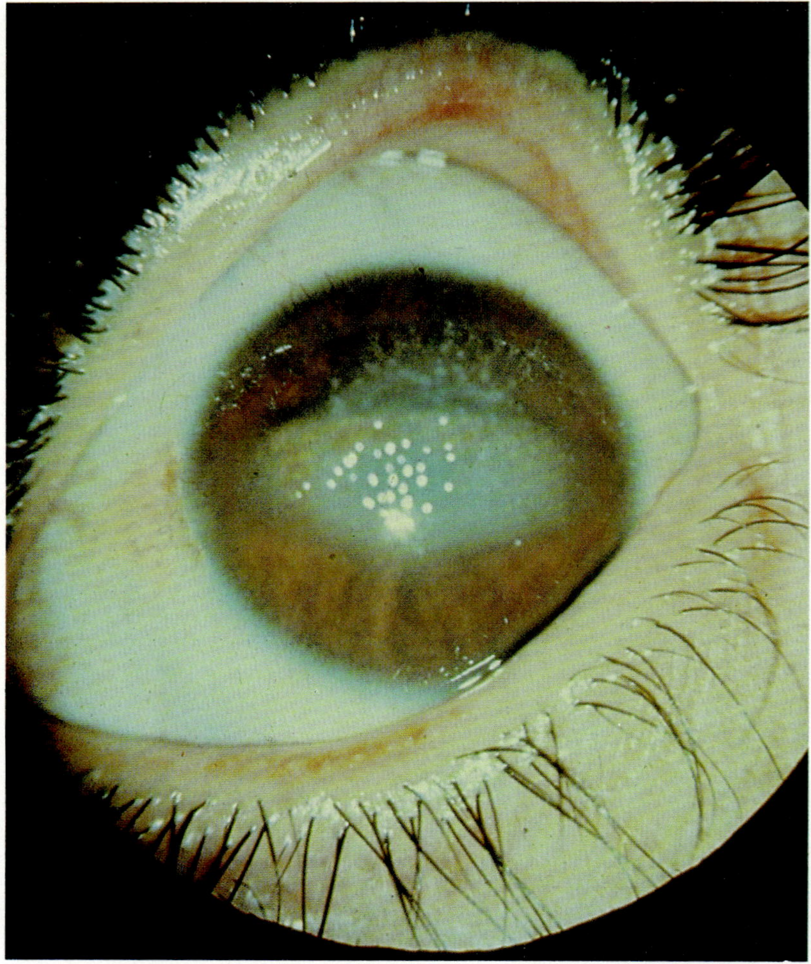

Figure 4.1. Eye with Labrador keratitis. (Courtesy of K. Kenyon, MD.)

amounts of UV irradiation overlap. However, a direct causal relationship between UV light and pterygium has not been proved.

Over the years the detectives of ophthalmology have unearthed
a considerable number of clues in the pterygium puzzle. It might
be instructive to consider some of the scattered pieces of evidence
to see if they fit into a convincing theory. The following are known:

1. Pterygium rarely develops before puberty or after the age of
   50.[17,18]

2. Pterygium is primarily found in hot environments, with the
   incidence rising toward the equator [17,18]; however, it also occurs
   in inhabitants of tropical rain forests, in indoor workers,[20] and
   in welders wearing UV-absorbing face shields.[21]

3. Histologically, the pterygium consists of a core of denatured
   collagen surrounded by lymphocytes and plasma cells.[22]

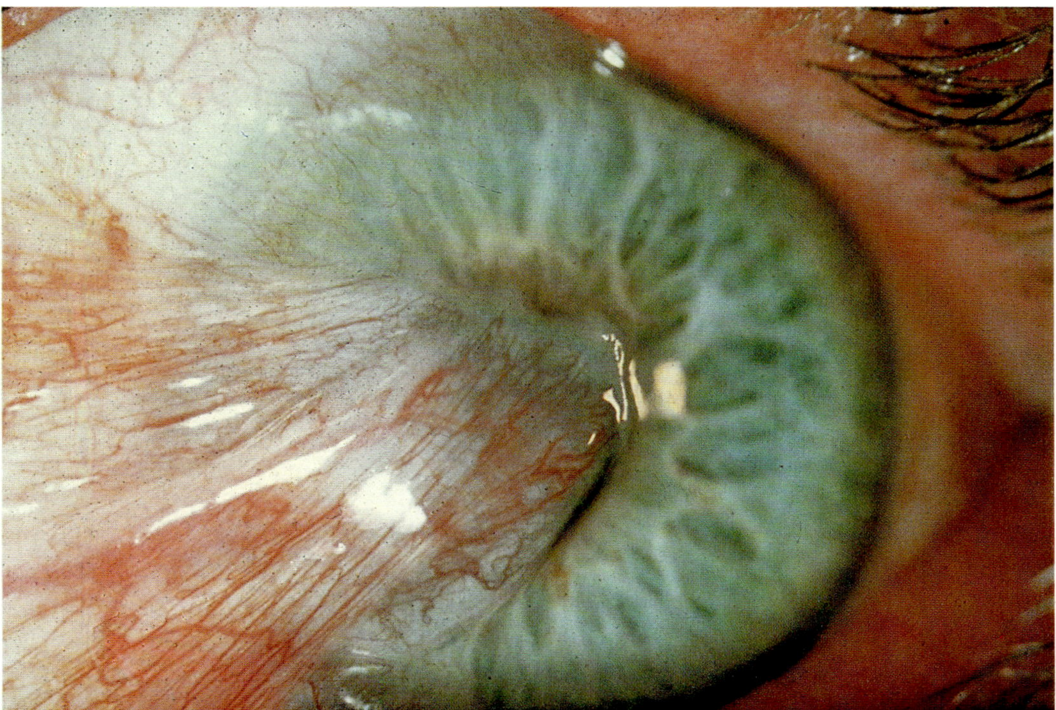

Figure 4.2. Eye with a pterygium.

4. Pterygium develops initially in the dominant eye. It has further been observed that the nondominant eye stays closed while the dominant eye remains open in a very bright environment.[23]
5. Nasal pterygia are much more common than temporal pterygia.[24]

These facts lead one to wonder if there might not be an endogenous substance produced by the body which:

1. Is present in higher amounts between puberty and the middle years
2. Is secreted when the environment is hot or when the face is enclosed
3. Is capable of denaturing conjunctival collagen
4. Commonly enters the eye via the nasal canthus
5. Cannot enter the closed eye

Since the forehead is the closest site capable of emptying large amounts of endogenous secretion under conditions of heat stimulation, we decided to apply a dye solution (rose bengal) to the forehead in the areas of sweat production and see where it flowed. Figure 4.3 demonstrates the progress of the dye as it flowed along the brow and down the side of the nose. Note that it entered

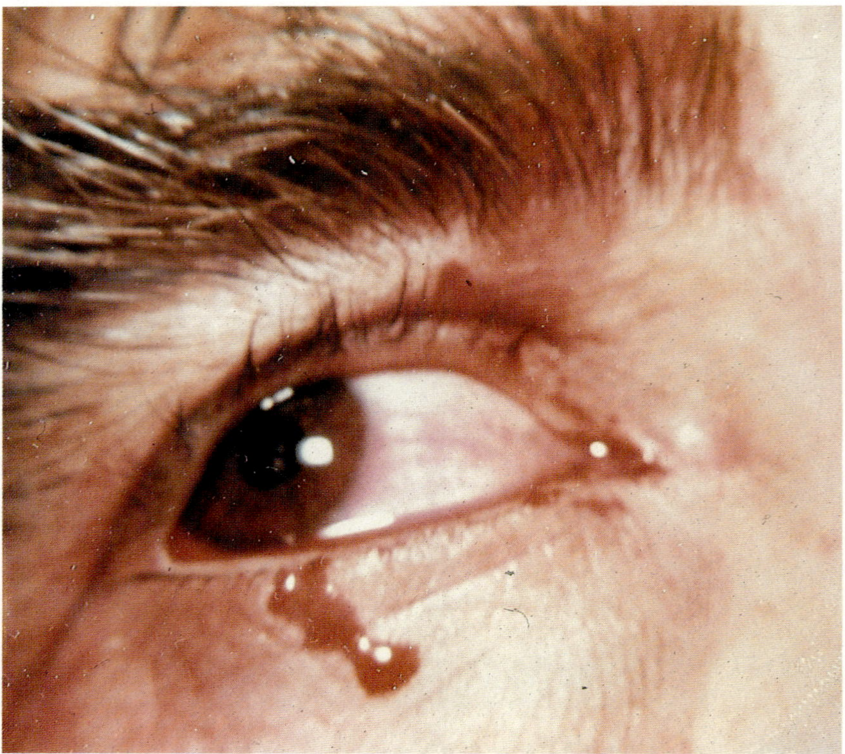

Figure 4.3. Path taken by the dye rose bengal as it left the forehead and trickled along the nose into the inner canthus.

the nasal canthus and collected in lines along the exposed nasal conjunctiva. Although very crude, these studies, plus the data collected from the literature, suggest that chronic stimulation of the conjunctiva by something present only in adult perspiration— perhaps lactic acid [25]—may slowly, unrelentingly, start the chain of events in pterygium production. Of course, variation of facial anatomy and ocular prominence could act to modulate the dose of acid perspiration that trickles from the forehead along side of the nose to enter the nasal canthus. Actually, with this hypothesis in mind, the moderate pterygium can be pictured as a protective plug that functions to keep further perspiration from entering an eye, while squinting in the bright sun. Does such a theory allow UV exposure to fit into the story of pterygium development? Two findings work against its playing a major role. In a sunny or reflective environment, the nose would prevent more nasal-directed light from striking the eye than temporally directed light. Yet, nasal pterygia are much more common than temporal pterygia. Then, too, pterygia occur in people working indoors [20] or in welders who wear protective masks that are specifically designed to prevent

UV light from striking the eye.[21] However, it is possible that once an elevated conjunctival lesion starts to form and a resulting dellen of the cornea develops, the limbal crevice that forms reflects concentrated levels of UV light onto the limba cornea, resulting in the unique corneal opacification.

It would be interesting to note the incidence of pterygia in diamond mine workers. These people work in very deep, very dark, and very hot environments where there is very little light of any kind.

Corneal Stromal Inlays

With interest in corneal refractive surgery heightening, one reads of the implantation of refractive elements within the corneal stroma. At present the high refractive index, polymer polysulfone is being tested.[26] Interestingly, certain forms of polysulfone are, in fact, used as UV sensors because they change characteristics under the influence of UV light.[27] These findings raise an important consideration in the development of stromal inlays. The ideal polymer, for such usage, should not be vulnerable to change under UV exposure.

Finally, a chapter on light-induced diseases of the conjunctiva and cornea would not be complete without a word about the effects of irritating air pollutants. Specifically, industrial and automobile exhausts belch hydrocarbons and nitric acid into the atmosphere. Ultraviolet radiation breaks some of these down into superoxide radicals, hydroxyl radicals, and hydroperoxyl radicals. These highly oxidizing free radicals react with organic pollutants to produce eye irritants such as formaldehyde (HCHO), peroxyacyl nitrate ($CH_3CO-O-O-NO_2$), and acrolein ($CH_2CH\ CHO$).[28]

References

1. Voerhoeff FH, Bell L, Walker CB: The pathological effects of radiant energy on the eye. An experimental investigation with a systemic review of the literature. Proc Am Acad Arts Sci 51:630–818, 1916.
2. Pitts DG, Tredici TJ: The effects of ultraviolet on the eye. Am Ind Hyg Assoc J 32:235–246, 1971.
3. Pitts DG: A comparative study of the effects of ultraviolet radiation on the eye. Am J Optom 50:535–546, 1970.
4. Cogan DG, Kinsey VE: Action spectrum of keratitis produced by ultraviolet radiation. Arch Ophthalmol 35:370, 1946.
5. Terrestrial solar spectral irradiance tables at air mass 1.5 for a 37 degree tilted surface. ASTM standard E892. Vol 12.02. American Society for Testing and Materials, Philadelphia, 1982.

6. Lerman S: Radiant Energy and the Eye. Macmillan Inc, New York, 1980.

7. Buschke W, Friedenwald JS, Moses SG: Effect of ultraviolet irradiation on corneal epithelium, mitosis, nuclear fragmentation, post traumatic cell movements, loss of tissue cohesion. J Cell Physiol 26:147, 1945.

8. Hemmingsen EA, Douglas EL: Ultraviolet radiation thresholds for corneal injury in antarctic and temperate zone animals. Comp Biochem Physiol 32:593–600, 1970.

9. Zuclich JA, Kurtin WE: Oxygen dependence of near ultraviolet induced corneal damage. Photochem Photobiol 25:133–135, 1973.

10. Zuclich JA: Ultraviolet induced damage in the primate cornea and retina. Curr Eye Res 3:27–34, 1984.

11. Brenner W, Grabner G: 3H-thymidine autoradiography of guinea pig cornea and skin after exposure to solar simulating radiation. Von Graefe Arch Klin Exp Ophthalmol 216(4):319–325, 1981.

12. Harm H: Damage and repair in mammalian cells after exposure to non-ionizing radiation. III: Ultraviolet and visible light irradiation of cells of placental mammals, including humans, and determination of photo repairable damage in vitro. Mutat Res 69(1):167–176, 1980.

13. Johnson GJ: Aetiology of spheroidal degeneration of the cornea in Labrador. Br J Ophthalmol 65(4):270–283, 1981.

14. Norn MS: Spheroid degeneration of cornea and conjunctiva. Prevalence among Eskimos in Greenland and Caucasians in Copenhagen. Acta Ophthalmol 56(4):551–562, 1978.

15. Freedman A: Labrador keratopathy. Arch Ophthalmol 74:198–202, 1965.

16. Johnson GJ, Overall M: Histology of spheroidal degeneration of the cornea in Labrador. Br J Ophthalmol 62:53–61, 1978.

17. Cameron EE: Pterygium Throughout the World. CC Thomas Publ, Springfield, IL, 1965.

18. Elliot R: The aetiology and pathology of pterygium. Trans Ophthalmol Soc Aust 25:71, 1967.

19. Moran DJ, Hollows FL: Pterygium and ultraviolet radiation: a positive correlation. Br J Ophthalmol 68:343–346, 1984.

20. Detels R, Dhir SP: Pterygium: geographic study. Arch Ophthalmol 78:485, 1967.

21. Karai I, Horiguchi S: Pterygium in welders. Br J Ophthalmol 68:347–349, 1984.

22. Hogan MJ, Alvarado J: Pterygium and pinguecula electron microscopic study. Arch Ophthalmol 78:485, 1967.

23. Jensen OL: Pterygium, the dominant eye and the habit of closing one eye in sunlight. Acta Ophthalmol 60(4):568–574, 1982.

24. Youngson RM: Pterygium in Israel. Am J Ophthalmol 74:954, 1972.

25. Corruthers C: Biochemistry of skin, in Health and Disease. CC Thomas, Springfield, IL, pp 90–152, 1962.

26. Choyce DP: Semirigid corneal inlays used in the management of albinsim, aniridia and ametropia, in Henkind P (ed). ACTA XXIV

Int Congress of Ophthalmology. JP Lippincott Co, Philadelphia, 1983.

27. Safran MJ, Rosenthal FS, Taylor HR: Measurement of ocular ultraviolet exposure in a anthropomorphic model. Invest Ophthalmol Vis Sci [Suppl] 26(3):213, March 1985.

28. Leighton P: Photochemistry of Air Pollution. Academic Press, New York, 1961.

5

Light Damage to the Lens

Seymour Zigman

Interaction of the Lens with Radiant Energy

Within the last ten years much effort in eye research has been devoted to the consideration that excessive exposure to environment radiant energy can damage the ocular tissues.[1-3] Prior to this time very few research efforts were devoted to this type of study, which has now permeated all areas of eye research. It is now accepted by most basic and clinical researchers that shorter wavelengths of light in the blue and near-ultraviolet (UV) range must be considered as additive factors in ocular tissue damage leading to reduced visual function.

The natural environment provides high irradiances of light in the short wavelength blue and long wavelength UV range. Whether or not these wavelengths of radiant energy can damage the ocular tissues depends on the irradiance that reaches the eye, the degree of transmission through the anterior ocular tissues, the threshold of energy required to induce a change, and the repair potential of the tissue of itself. Because of the presence of ozone in the atmosphere, only radiant energy longer than 295 nm reaches the earth and will be considered here.

A generalization to allow discussion of different wavelength ranges of UV light and their ability to alter ocular tissue follows. Ultraviolet-A is defined to include radiant energy in the wavelength range of 315 to 400 nm; UV-B includes the wavelength range from 315 down to 280 nm; and UV-C represents UV radiation of wavelengths shorter than 280 nm. In a mechanistic sense, it has been found that UV-A exerts its influence on biochemical

Table 5.1. Transmittance of near-UV radiation through the cornea and lens.

Tissue	Wavelength (nm)				
	300	320	350	380	360
	Percent Transmittance				
Cornea	8	60	70	78	75
Lens	0	2	3	4	—
% reaching the retina	0	1.2	2.1	3.1	25

systems by photosensitized effects, whereas UV-B can directly alter the macromolecules with biological functions (i.e., proteins and nucleic acids). Ultraviolet-C is essentially the germicidal wavelength range that kills cells by damaging their DNA and other nucleic acid consitutents directly via thymine dimers and chain breaks. Only UV-A and UV-B need be considered since UV-C does not penetrate the ozone layer to reach the earth. Ultraviolet-C is practically provided only as an emission from certain artificial lamps. It is used for sterilization, analysis, and certain medical treatments.

For orientation, the penetration of radiant energy through the ocular tissues is illustrated in Table 5.1. Ultraviolet-B is maximally absorbed by the cornea, while UV-A penetrates so as to be maximally absorbed by the lens. Ultraviolet-B has been implicated as the wavelength range responsible for sunburn and photokeratitis. Very little UV energy reaches the retina of normal eyes. However, appreciable irradiances of UV-A and some UV-B do reach the retina in aphakic or lensless eyes. Ultraviolet-A radiant energy is represented in natural sunlight at a much higher irradiance than is UV-B, and is sufficient to interfere with the growth and differentiation of lens epithelium and with photoreceptor maintenance in the retinas of aphakic eyes.

The threshold for irradiances in the different wavelength ranges at which ocular tissue damage occurs is not known precisely. However, Table 5.2 provides a short summary of the available ocular tissue threshold values. Comparatively speaking, the threshold for UV-B radiant energy damage to the cornea is approximately one tenth of that for damage to the internal ocular tissues (i.e., lens and retina), whereas the corneal threshold for UV-A energy is only one half that for lens damage.[4] Using krypton laser exposure, whose wavelength is intermediate between UV-A and UV-B, Zuclich and Connaly[5] showed that in primates the retinal threshold for damage is ten times that of the corneal threshold. Since the

Table 5.2. Comparative threshold for ocular tissue damage from near-UV radiation.

Wavelength (nm)	Cornea	Lens	Retina
		(J/cm^2)	
300	0.015	0.15	0.185
315	7.3	4.5	—
320	7.5	12.6	—
325	18.0	50.0	5.0
335	11.0	>15.0	—
365	43.0	>70.0	4.3

shortest UV-wavelengths that are absorbed by the lens damage it most, corneal absorption has a great influence on whether or not the lens will receive energy capable of causing it harm. The wavelength range that is realistically the most hazardous to the lens is between 310 and 390 nm. Of course, within this range the shorter wavelengths that reach the lens are the most hazardous to it.

Little information is available to relate the repair potential of ocular tissues for damage due to UV-light radiation. Practically speaking, repair of corneal epithelial cells takes place readily, as is shown by the rapid regrowth of corneal epithelial cells after the insult of corneal photokeratitis. The neural origin and characteristics of the cells of the retina (including both neural cells and photoreceptors) do not support the possibility that appreciable repair could take place. The special types of growth and of differentiation of the lens epithelium also make repair a remote possibility. However, it has been shown by Jose and Yielding[6] that unscheduled DNA synthesis does occur in the lenses of animals exposed to UV-B radiation. Owing to the conservative growth mechanism of the lens, damage due to an acute exposure to UV-A or B (as with laser) could conceivably not ever appear in a practical sense. A small point of peripheral lens damage would be relocated internally and perhaps would never appear in the visual axis. Chronic exposure, however, would lead to lens opacity owing to the long-standing damage to the epithelial cell growth and differentiation process. The latter type of opacity has actually been found to occur in mice.[7]

While clear-cut data on the damage to the lens by UV-radiation are obtainable in animals, UV-radiant energy is also a factor in human cataract formation, as shown by epidemiology studies.[2] Often these studies are retrospective and rely on the accumulation of data not specifically designed to show a relationship between

UV exposure of humans to sunlight and the formation of cataract. Other problems include the fact that experimental designs vary, and confounding variables differ from one study to another. The levels of UV-B in the environment are associated with human cataract in a positive manner, as shown by the study of Hiller et al.[8] Brilliant[9] showed that sunlight exposure per day was a positive factor in the development of cataracts in an area of Nepal, India, while altitude was not a factor. Taylor's studies[10] brought up a statistically significant relationship between the exposure to sunlight-UV and cataract in humans. All of the epidemiological studies investigating the relationship between UV exposure in sunlight and cataract have been positive.

Basic Studies of Ocular Damage Due to UV-Exposure

There are few molecular species to serve as targets for UV-A damage to biological and ocular systems. For the most part, tryptophan is the most prominent target for near-UV damage. Although the absorption maximum of tryptophan is in the 280 nm range, there is a long tail of absorption into the UV-A region. Other absorbers of UV-A include the yellow pigments that are present in the lenses of most diurnal animals and man. These pigments can be of a low molecular weight, water-soluble nature, or of a macromolecular nature being covalently linked to proteins and membrane systems. It appears that the proteins of the lens are susceptible to photochemical change induced by their exposure to UV-A and -B either directly, or by virtue of photosensitizers. Changes in the protein aggregation and other major changes in the crystallins are well documented below. For example, it has recently been shown that many enzymes that are involved in the control of oxidation reactions in the lens and in osmoregulatory reactions lose appreciable catalytic activity upon UV exposure. These changes damage the lens and lead to cataract. Table 5.3 indicates the action of UV-A on Na/K adenosine triphosphatase (ATPase) of the lens. Numerous other enzymes have been found to be sensitive to UV-A in the presence of a sensitizer such as tryptophan.[1,2]

The classical experiment in which large numbers of animals were exposed to chronic ambient near-UV exposure (mainly UV-A) at subsolar irradiances was reported approximately 12 years ago.[7] Chronic exposure to near-UV illumination with peak emission at 365 nm and an irradiance of 500 mW/cm^2 led to abnormal lens epithelial cell differentiation and enhancement of aggregated proteins in the lenses of mice after several months. Cortical cataracts

Table 5.3. Effect of in vitro near-UV radiation at 5 mW/cm^2 and 360 nm on Na$^+$/K$^+$ ATPase activity of rat lens epithelium plus cortex.

Irradiation time (of hours)	Na/K ATPase activity (% of control)
5	60
10	40
15	22
20	8

developed after a year and a half. No corneal damage was observed throughout the course of this experiment. However, within 6 weeks of this type of ambient exposure to UV-A radiation, the retina exhibited the following damage: thinning of the photoreceptors, wandering macrophages in the outer segments, and eventually loss of all photoreceptors. When black pigmented animals were used instead of the white albinos mentioned above, none of these changes occurred. Thus, iris melanin seems to be capable of protecting sensitive lens epithelium from UV-A damage. Exposure of albino mice separately to UV-A and UV-B wavelengths indicated UV-B damage to cornea and UV-A damage to lens and retina, which findings completed the proof that UV-A leads to changes mainly in the lens and retina, whereas UV-B leads to corneal keratitis.[11]

Recent research has employed the diurnal grey squirrel as an experimental animal for studies of near-UV interaction with ocular tissues. A prominent feature of the lenses of these squirrels is that they contain a yellow near-UV absorbing pigment similar to that present in very young human lenses. Besides this feature, the lenses are quite large, so as to be suitable for biochemical studies. Using this animal, it is not difficult to remove the lens and to show that the lens serves as a near-UV filter. The retina, which does not receive such light in the normal intact eye, is damaged by it in the aphakic eye.[12]

In these experiments grey squirrels were exposed to 40 W black light bulb (BLB) lamps for 12 hours per day, which provided approximately 1.5 to 2 MW/cm^2 of radiant energy at 365 nm. Control room light and UV-exposed animals were maintained for periods up to 2 years. Periodically, the animals were sacrificed and the lenses dissected in several ways. Samples of small portions of these lenses were extracted by homogenization; low molecular weight soluble proteins and high molecular weight insoluble proteins were separated by centrifugation. The tissue extracts obtained in this way were subjected to high-pressure liquid chromatography (HPLC) on gel filtration Toya Soda Kiselgur (TSK) columns. When the most anterior and posterior polar portions of the lenses of grey squirrels were examined, it was found that major changes in the distribution of the soluble protein bands were observed in the anterior cortical portion of the lens; however, little change in the posterior cortical area was observed as compared with a control cortex sample. The changes induced by UV light exposure were: 1) losses of the heaviest or void volume proteins, 2) losses of pro-

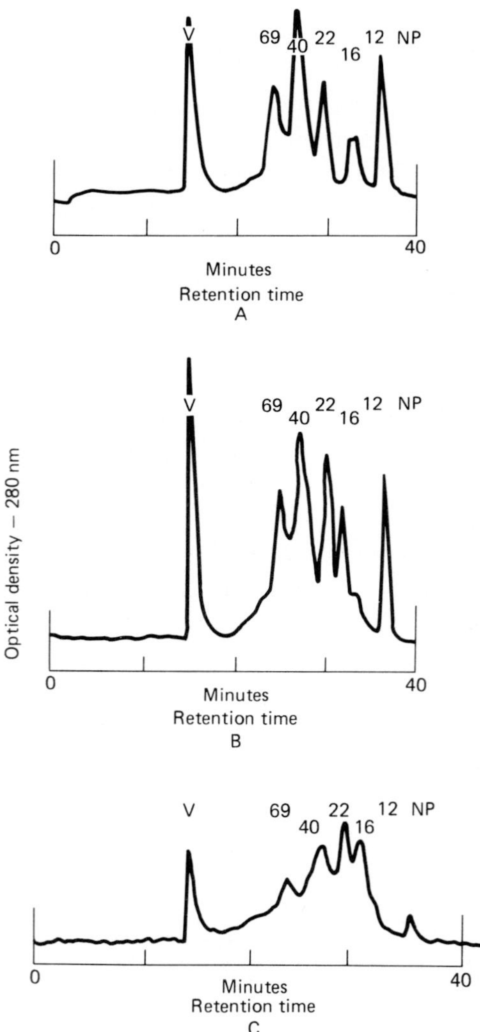

Figure 5.1. Gel permeation HPLC profiles of grey squirrel lens outer cortical soluble proteins. A = dark controls; B = in vivo exposed to near-UV radiation for 7 months; C = in vitro exposed to near-UV radiation for 18 hours.

teins with intermediate molecular weights (69,000, 40,000, 22,000 d), and 3) increases in peptides with molecular weights of 16,000 d (see Fig. 5.1). Also observed was a marked increase in the aggregation of both soluble and insoluble protein fractions in the UV-exposed lens cortex. These findings indicate that both aggregation of soluble proteins and enhancement of low molecular weight peptides result from direct near-UV exposure in vivo. Since near-UV radiation enters the lens anteriorly, where it is maximally absorbed by the protein and pigment present in the lens, much of the incident energy has been absorbed before it can reach the posterior of the lens. Thus, it appears that the anterior cortical lens soluble proteins are the most susceptible to photochemical changes induced by near-UV radiation.

In vitro studies and chemical studies have supported and enhanced the findings derived from in vivo studies. When water-soluble extracts of concentric layers of squirrel lenses were subjected to near-UV radiation (Woods lamp; 5 mW/cm^2), a precipitate readily formed within a few hours. After removal of the 13,000 g precipitate by centrifugation, both soluble and insoluble portions of the extract were examined. The results of this experiment is shown in Figures 5.2a–d, which show that in all layers (i.e., outer, middle, and inner), the level of soluble protein was diminished and the level of insoluble protein was enhanced owing to near-UV exposure. When the solubility characteristics of the insoluble precipitate were examined it was found that there was an enhancement of the urea-soluble, the sodium dodecyl sulfate (SDS)-soluble, and SDS + dithiothreitol (DTT)-soluble portions of this aggregate.[13]

In vitro changes in the soluble proteins themselves, as studied by HPLC, revealed similar changes to those found in vivo. That is, losses in the heaviest void volume protein, enhancement of aggregated insoluble profiles, and increase in lower molecular weight peptides were observed. These changes were prevalent in the outermost (cortical) and second layers of the lens but changed little in the nucleus. An attempt was made to elucidate the mechanism whereby this aggregation occurred by observing the changes in fluorescence properties of the soluble and insoluble proteins that were altered. There appeared to be a loss of the usual protein (tryptophan) fluorescence (i.e., 290 excitation and 330 nm emission) and the production of a new, longer wavelength fluorescence (i.e., 343 excitation and emission at 388 nm). Changes in the lens-soluble protein fluorescence is illustrated in Figure 5.3. The above described changes in protein distribution and fluorescence were much diminished when the water-soluble, low molecular weight, lens pigment (identified as *N*-acetyl 3-hydroxykynurenine in the

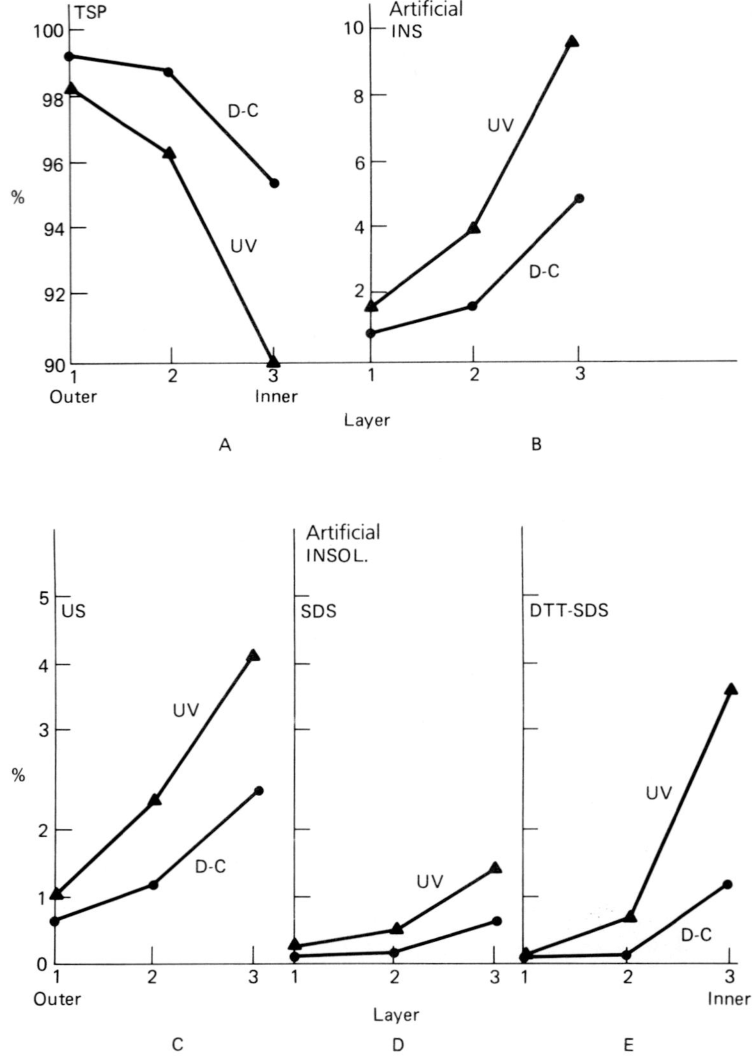

Figure 5.2. Phase shift of grey squirrel lens layer soluble proteins into aggregated species (INS) due to exposure of extracted total soluble protein (TSP) to near-UV radiation. DC = dark control; UV = UV exposed; US = portion of aggregate solubilized in 8 M urea; SDS = portion of the aggregate soluble in 1% SDS plus 50 mmol/L dithiothreitol.

squirrel lens) was removed from the protein by pressure dialysis. Protein distribution changes and fluorescent changes occurred to a much greater extent when the low molecular weight, water-soluble pigment, was left in the soluble phase during irradiation. Polyacrylamide gel electrophoresis was used to observe the aggregation of the formerly soluble proteins due to near-UV exposure in both the presence and the absence of this chromophore (Fig.

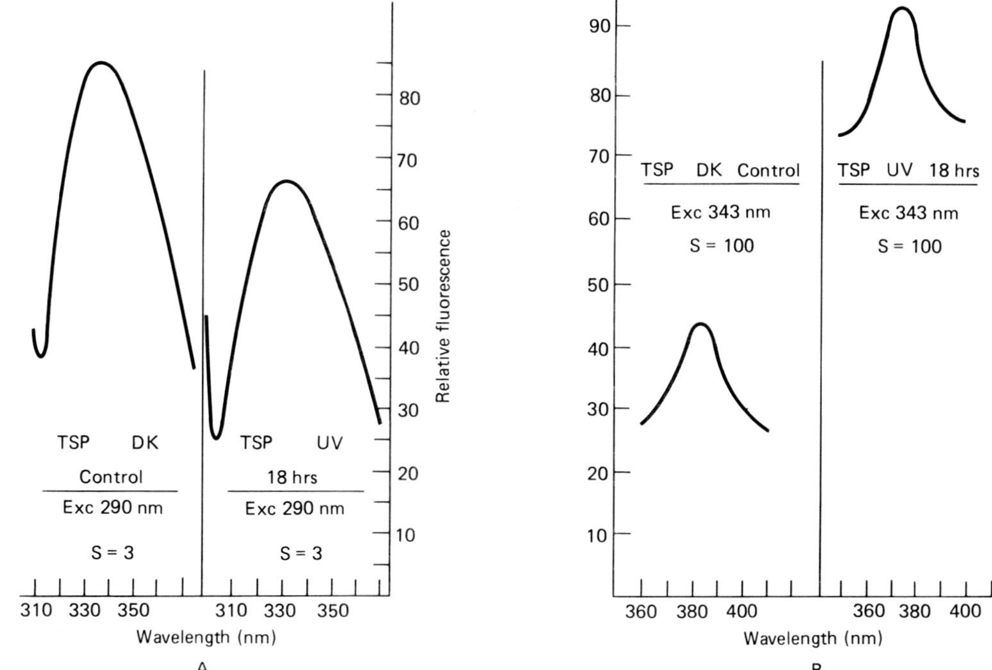

Figure 5.3. Enhancement of non-tryptophan fluorescence (b) and decrease of the usual tryptophan fluorescence (a) of grey squirrel lens total soluble proteins (TSP) due to in vitro exposure of TSP to near-UV radiation for 18 hours. a) fluorescence is excited at 290 nm and emits at 330 nm; b) fluorescence is excited at 343 nm and emits at 383 nm.

5.4). The degree of aggregated material observed at the top of the gels is much greater in the presence of the chromophore than in its absence.

Another mechanism suggested for the aggregation of proteins due to near-UV exposure of the lens has been made by Goosey et al,[14] whose data indicated that singlet oxygen formation was a major mechanism whereby proteins of the lens were aggregated by near-UV light exposure. By including a well-known inhibitor of singlet oxygen formation (sodium azide), it was shown that aggregation of lens proteins in the presence of the chromophore by near-UV exposure is totally inhibited when sodium azide is present. This indicates that singlet oxygen formation has an important role in causing the aggregation of squirrel lens proteins owing to near-UV exposure. The importance of the low molecular weight UV-absorbing chromophores as sensitizers is further emphasized by these findings. Singlet oxygen may only be one of the oxidants leading to protein aggregation and other changes observed.

The findings and projected mechanisms whereby near-UV light

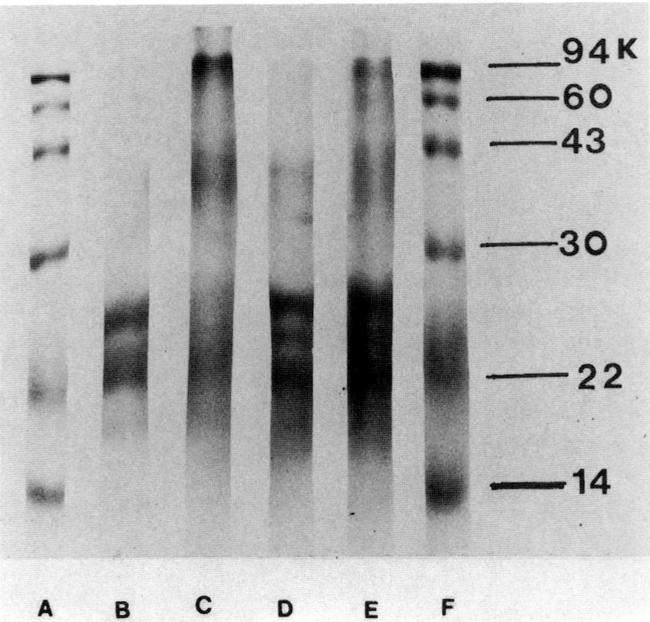

Figure 5.4. Near-UV radiation-induced aggregation of squirrel lens soluble proteins as influenced by the presence and absence of lens pigments.
A, F = standard proteins; B = dark control plus pigment; D = dark
control minus pigment; C = UV exposed plus pigment; and E = UV
exposed minus pigment.

alters the squirrel lens can be applied to the changes found in
certain types of human cataract. For example, UV exposure has
been implicated in influencing the production of the brunescent
cataract. In this type of cataract there is an enhancement of aggregated protein, a change in the protein fluorescence to longer wavelengths, and an increase of peptide material, most probably associated with proteolytic change. The insoluble proteins of these types
of cataracts are known to be cross-linked by fluorescent cross-links.
The cross-links of insoluble proteins derived from brunescent cataracts have been identified by Dillon, et al[15] as being tryptophan
oxidation products.

The above description of experimental findings in animal and
human studies supports the concept that near-UV light in the
environment is absorbed by the lens, alters lens proteins, and
thereby contributes to cataract formation.

Recent work using squirrels has shown that removal of the lens
allows sufficient ambient near-UV radiation to reach the retina to
do it harm. When the lenses of squirrel eyes were removed, and
the animals were then maintained under ambient near-UV radiation as described above, the photoreceptors of the retina were
damaged readily in relatively short times.[12] This proves that the

lens, with its high concentration of pigments and proteins, filters near-UV energy entering the eye before it can reach the retina, where it can damage the photoreceptors.

Application to Eye Protection and Practical Considerations

It is important to apply the above-stated information to a discussion of the need to reduce or eliminate the entry of near-UV radiation into the eye. Nature has provided a yellow filter in the ocular lens of diurnal vertebrates to absorb near-UV energy so as to improve visual acuity by eliminating chromatic aberration and glare.[16] However, another function of lens pigment has now been ascribed, at least in squirrels and primates, to retinal protection against actual near-UV-induced damage to the photoreceptors.[12] But in the process of absorption, the lens pigments serve as photosensitizers that enhance photooxidative changes in lens structural proteins and diminish certain essential enzymatic activities. Such changes result in light absorption and scattering in the lens, which contribute to cataract formation. The ultimate degree of such photosensitized enhancement of scattering and absorption is seen in deeply pigmented human cataracts. Such pigmentation does not protect the lens from other varieties of cataracts, which result from abnormal epithelial cell differentiation due to interference with protein and nucleic acid metabolism in these actively growing and dividing cells. Exposure of these active lens epithelial cells is minimized by the presence of a melanin-lined iris and the pupil size. Regardless of which of the specific mechanisms is involved, it is generally agreed that the elimination of the near-UV wavelength from the eye is a worthwhile undertaking, as based on both basic and epidemiological grounds. The most practical application of this information for eye health care is the use of UV-filtering spectacles, contact lenses, and intraocular lens implants in environmental conditions providing high irradiances of near-UV radiation.

Yellow (including green) or amber-tinted sunglasses that filter out radiant energy of wavelengths shorter than 400 nm are to be recommended. Such filtration eliminates the wavelengths that are not useful for vision and have the greatest potential to cause photochemical changes in the lens. While neutral density filtration diminishes visible light wavelengths equivalently, it may not diminish UV transmission as well. Since the pupillary response is to visible light, a dilation of the pupil may admit a greater level of UV-radiation to impinge upon the more sensitive growing and dividing lateral lens epithelial cells. By calculation, a dilation of the pupil from 3 to 7 mm exposes ten times the area of the underlying ocular lens.

It is not always possible to determine how effective sunglass lenses are in filtering near-UV wavelengths. The public must rely on information pertaining to the efficacy of sunglass lenses or near-UV filters provided by manufacturers. Indeed, spectacles labeled "sunglasses" are not always effective near-UV absorbers and may even transmit near-UV energy more intensely than visible light. Proposals have often been made to require transmission data to be provided with all sunglasses.

The importance of protecting the retina from continuous or repetitive exposures to blue and near-UV light has recently been emphasized. Solar retinitis and photic maculopathies are now believed to result from the exposure of humans to intense short wavelength light.[17,18] These types of damage to the retina reduce the efficacy of the photoreceptors in the visual process and if repeated frequently can result in total degeneration of many of the photoreceptors of the retina.[12] Of major concern are the eyes of aphakic individuals whose natural ocular lenses have been removed, since standard replacement artificial lenses of polymethylmethacrylate do not absorb the UV and blue visible light before it reaches the retina. New UV-absorbing intraocular lenses are presently becoming available for investigational use.

How much UV light should be absorbed by sunglasses and other eye-protective devices? The American National Standards Institute and American Industrial and Governmental Hygienists have provided a threshold for eye damage of 1 mW/cm^2 of near UV-light (320–400 nm) for periods greater than 10^3 seconds (16 minutes), and energy 1 J/cm^2 over the same time period. On a sunny day in the northern hemisphere at midday in summer, the sun without clouds provides from 5 to 10 mW/cm^2 in this wavelength range. It is obvious that long-term direct exposure of the eye to unfiltered sunlight provides a dose of long wavelength UV that exceeds this threshold if continued for as long as an hour. Sunglasses should reduce the level of near-UV to about 0.1 mW/cm^2. For this to occur, the near-UV irradiance must be cut by 90%.

Since the UV representation is greater at lower latitudes and higher altitudes, the degree of absorption by sunglasses used in these areas should be greater. Areas that are snow covered, sandy areas, and the water reflect much of the near-UV energy.

Special circumstances in which total absorption of long wavelength UV and short wavelength blue visible light is needed are when humans are under treatment with photosensitizing drugs (i.e., psoralen, epinephrine, primaquine) and in those humans who no longer have a natural ocular lens (i.e., aphakic eyes after surgery for cataracts). The damage thresholds for the retina in aphakia are about one tenth lower. It has recently been shown that only one tenth of the energy that is required to produce a retinal defect in a monkey eye at 500 nm is required to produce similar defect at 450 nm.[19] This means that without lens absorption, the retinas of monkeys and, as stated previously, of squirrels can be damaged by near-UV radiation even more readily. This still needs to be proven in humans. Aphakic individuals should be provided with artificial lenses that totally absorb the short and the long wavelength UV energy (below 400 nm).

Conclusions

By far, the most convenient and effective means to protect the eye from potentially hazardous environmental radiant energy is to use artificial lenses that absorb this energy. Although there are no ideal sunglass lenses, important functions of such lenses must be not only to reduce the total visible light level so as to eliminate glare, chromatic aberration, and lens fluorescence, but also to protect the ocular tissues from photo-induced damage. The accomplishment of these functions is especially important under special circumstances. These include human outdoor activity in locations of low latitude and intense reflectance when such exposure is repetitive over many years; use of photosensitizing medications, and indoor exposure to the high UV irradiances emitted by industrial, therapeutic, and cosmetic light sources.

Not all sunglasses accomplish the goal of eliminating long wavelength UV radiation from entering the eye. In some, there are actually UV-transmitting windows. In others, the lenses absorb enough visible light to cause pupillary dilation. However, if these lenses do not absorb an equivalent amount of UV radiation the ocular lens may be damaged.

Thus it is important that sunglass lenses for eye protection should prevent all near-UV energy from entering the eye, especially when repetitive exposure to bright direct or reflected sunlight is anticipated. The feature of sunglass lenses relative to the absorption of harmful wavelengths of light should be made very clear to suppliers and users of these devices.

References

1. Zigman S: Photochemical Mechanisms in Cataract Formation. *In* Mechanisms of Cataract Formation in The Human Lens, Duncan G (ed). Academic Press, New York, pp 117–149, 1981.

2. Zigman S: Recent research on near-UV radiation and the eye, in The Biological Effects of UV Radiation, Urbach F (ed). Preger Sci, NJ, 1986.
3. Lerman S: Radiant Energy and The Eye. McGraw-Hill, 1980.
4. Pitts D: Effects of UV Radiation in The Eye in Biological Effects of UV Radiation. Parrish, Anderson, Urbach, Pitts (eds). Plenum Press, New York, pp 177–215, 1978.
5. Zuclich JA, Connaly JS: Ocular damage induced by near-UV lasar radiation. Invest Ophthalmol 15:760–776, 1976.
6. Jose JG, Yielding KL: Photosensitive cataractogens, chlorpromazine, cause DNA repair synthesis in lens epithelial cells. Invest Ophthalmol Vis Sci 17:689–691, 1978.
7. Zigman S, Yulo T, Schultz J: Cataract indication in mice exposed to near-UV light. Ophthalmol Res 6:259–270, 1974.
8. Hiller R, Sperduto RD, Ederer F: Epidemiologic associations with cataract ion. The 1971–1972 Natural Health and Nutrition Examination Survey. Am J Epidemiol 118:239–249, 1983.
9. Brilliant LB, Grosset NC, Ram PP, Kolstad A, Lepkowski JM, Brilliant GE, Hawks WN, Parajasegaram: Association among cataract prevalence, sunlight hours, and altitude in the Himalayas. Am J Epidemiol 118:250–264, 1983.
10. Taylor H: The environment and the lens. Br J Ophthalmol 64:303–310, 1980.
11. Zigman S: The role of sunlight in human cataract formation. Sur Ophthalmol 27:317–326, 1983.
12. Collier R, Waldron W, Merrill D, Zigman S: Effects of ambient near UV-exposure on the aphakic squirrel retina. Invest Ophthalmol Vis Sci [suppl] 25:18, 1984.
13. Zigman S, Paxhia T, Waldron W: Effects of Near-UV Radiation on the Proteins of the Grey Squirrel Lens. Curr Eye Res 1987 (submitted).
14. Goosey JD, Zigler JS Jr, Kinoshita JH: Crosslinking of lens crystallins in a photodynamic system. Science 208:1278–1280, 1980.
15. Dillon J, Spector A, Nakanishi K: Identification of beta carbolines isolated from fluorescent human lens proteins. Nature 259:422–423, 1976.
16. Jacobs GH: Comparative Color Vision. Academic Press, New York, London, pp 39–45, 1981.
17. Ham WT Jr, Mueller HA, Sliney DH: Retinal sensitivity to damage from short wavelength light. Nature 260:153–155, 1976.
18. Mainster MA: Spectral transmittance of intraocular lenses and retinal damage from intense light sources. Am J Ophthalmol 85:167–170, 1978.
19. Ham WT Jr, Mueller HA, Ruffolo, JJ Jr, Clarke AM: Sensitivity of the retina to radiation damage as a function of wavelength. Photochem Photobiol 29:735–743, 1979.

6

Phototoxic Changes in the Retina

John Weiter

History of Retinal Light Damage

Throughout history, mankind has revered the sun as a source of power and well-being. The ancient Egyptians considered Ra, the sun god, as the supreme deity. The Greek sun god, Apollo, has a special place in medicine since his son Aesculapius, was considered the first physician, thus directly linking the power and the healing aspects of the sun. This common thread associating the sun with power and healing has been part of our civilization from time eternal.

The healing properties of sunlight were only therapeutically applied on a scientific basis in the past century. In 1896 in Copenhagen, Niels Ryberg Finsen opened his world-famous institute for light therapy. His work, mainly on phototherapy of skin lesions such as lupus vulgaris was awarded the Nobel Prize in 1903. His colleague, Johan Widmark, the first professor of ophthalmology at the Karoline Institute at Stockholm, applied Finsen's concepts specifically to the eye. Although ocular phototherapy was quite popular for the first several decades of this century, more effective methods of treatment have outdated the technique and it is now of historic interest only.

Only recently have the harmful aspects of light become apparent. The discomforture associated with looking at the sun and the recognition of the associated scotoma led to the possibility of light-induced retinal damage. Socrates advised that a solar eclipse should be observed only by looking at its reflection in water. Both Constantine VII and Galileo injured their eyes looking at the sun.[1]

For years the clinical interest focused on eclipse-related solar burns. Experimental retinal light damage dates from the pioneering work of Czerny[1] in 1867, in which chorioretinal burns were created by focusing sunlight in the eyes of animals. Deutschmann (1882) used focused sunlight and Widmark (1893) used a carbon arc light source to produce similar chorioretinal burns. Birch-Hirschfeld (1904) first showed that ultraviolet (UV) light could cause retinal damage at exposure levels below those necessary for thermal damage. Widmark (1891) suggested that aphakic erythropsia was caused by UV light on the retina that had previously been blocked by the lens. Van der Hoeve[2] in 1919 suggested that chronic exposure during ones lifetime to UV light contributed to both macular degeneration and cataracts.

Maggiore (1933) was the first to study experimentally induced chorioretinal burn in humans. These burns were created prior to enucleation for malignant melanoma using sunlight and a nitra lamp.[1] Meyer-Schwickerath[3] was the first to successfully employ photocoagulation in a clinical situation. In 1949 he first used focused sunlight to photocoagulate a retinal break. With the advent of artificial sources of intense light, such as arc lamps and lasers, the field of therapeutic photocoagulation has advanced rapidly. So, too, has our recognition of the possibility of clinically induced light damage.

Light from an indirect ophthalmoscope can cause irreversible retinal damage after only 15 minutes exposure in the monkey.[4] In contrast to photocoagulation, the damage does not occur immediately, rather it takes several days.[5] Similar to photocoagulation burns, the damage is greatest at the level of the photoreceptors and retinal pigment epithelium (RPE). Gross macular edema was often present early, but tended to disappear. Because of the possible relation to aphakic cystoid macular edema, retinal edema from light exposure from the operating microscope has received much attention recently. The operating microscope has been shown to cause macular lesions in both humans and primates.[6,7] This is not surprising, since the light source in popular types of operating microscopes can expose the patient's retina to 100 to 970 mW/cm^2 if the media is clear.[8] Although these lesions are not similar to cystoid macular edema (CME), it had been postulated that UV light from the operating microscope might be related to CME. A controlled study using a UV filter on the operating microscope failed to show a difference in the angiographic incidence of CME or in the visual outcome in patients undergoing extracapsular cataract surgery with a posterior chamber lens insertion.[9]

Although short-term UV exposure from the operating microscope does not appear to be a factor in the incidence of aphakic

CME, long-term postoperative UV light exposure may play a role. To test this hypothesis, a controlled study was performed,[10] comparing the incidence of CME in pseudophakic patients who received an intraocular lens (IOL) containing a UV-filtering chromophore as compared with pseudophakic patients who received an identical IOL without such a chromophore. The presence of a UV-filtering chromophore statistically reduced the incidence of fluorescein angiographically detectable CME. However, the presence or absence of the UV-filtering chromophore did not significantly affect visual acuity in the early postoperative period. It is possible that continued UV exposure over a longer study period would demonstrate an effect on vision, either from macular edema or other degenerative changes.

Mechanisms of Light Damage

Light can damage the retina in at least three fundamental ways: (1) photochemical, (2) thermal, and (3) mechanical. These types of damage are not mutually exclusive and areas of overlap occur between them. Mechanical damage results from high irradiance levels and short exposure times such that sonic transients or shock waves are created. Q-switched and mode-locked YAG photodisruptors are clinical instruments that use this form of light damage. Technology has only recently introduced this form of light damage, and its uses and hazards are being explored. Thermal damage results when enough light is absorbed to raise the temperature in the retina by 10° to 20°C. Clinicians have been familiar with this mode of retinal damage through the use of the xenon-arc and laser photocoagulators. An example of the overlap between mechanical and thermal light damage is seen during laser photocoagulation, when enough light is absorbed in the retinal pigment epithelium and surrounding tissue to cause a vapor bubble (or an "explosion") to occur.

Photochemical damage is induced by chemical reactions initiated by light. Unlike the high-power density and short exposure times that characterize both thermal and mechanical light damage, photochemical damage depends on prolonged exposure time to the more energetic short wavelengths such as blue and UV light. We previously demonstrated that UV light induces free radicals in the human lens.[11,12] The possibility of nonthermal damage in the retina from visible light was neglected until recently when Noell et al noted electroretinographic and histologic changes in rats exposed to ordinary fluorescent lights presented through a green filter for 24 hours.[13] These investigators noted that the retinal irradiance was far too low for a thermal lesion and postulated a direct light toxicity to the photoreceptors. They also noted a cumulative effect and a higher damage threshold for more pigmented animals. These nonthermal photochemical reactions are more and more recognized as the most probable causes of hazards of ophthalmic instrumentation[14] such as the indirect ophthalmoscope, as noted above. Solar retinopathy and welding arc maculopathy most likely also have a photochemical component. Ophthalmologists need to be aware of photochemical light damage since it is produced by absorbed energy levels far below those that are required to produce thermal damage. For this reason its properties merit consideration.

Intraocular Microenvironment

Retinal Irradiance

One of the most enduring and widely believed concepts is that the macula receives a greater irradiance than the rest of the retina. This belief is related to the idea that the biological optics of the eye focuses light in the foveal-macular region, therefore resulting in a greater irradiance. This increased macular light exposure is often stated as contributing to macular degeneration.[15] Although the ocular optics would focus light on the fovea when gazing at a light source (such as solar retinopathy during eclipse observation), there is scant evidence that a photochemical lesion would be more severe in the macula when exposed to diffuse light. Young[15] measured the luminance of the retinal image of a small light source from the back of the excised rat eye as a function of eccentricity and noted decreased luminance. Unfortunately, the quality of the imaging system deteriorates with peripheral angle, and therefore the luminance of a small image decreases as the light is spread over a larger area. Calculations of retinal irradiance with Ganzfeld (diffuse full field) illumination in a wide-angle theoretical model eye has shown that the retinal light distribution is nearly homogeneous over the whole retina.[16] The homogeneity was little influenced by the size of the pupil or the shape of the optical surfaces. Similar studies and calculations by Bedell and Katz[17] and Pomerantzeff and Pflibsen from our laboratory (personal communication—to be published) have shown that the retinal illuminance of a diffuse light source should remain essentially constant at the posterior pole of the eye (up to approximately an 80° field). The difference between these two schools of thought is related to the light source being evaluated. Indeed, the image of a small light source does deteriorate in the eye as a function of eccentricity and the corresponding retinal irradiance decreases as the light spreads. A diffuse light source can be thought of as many light sources side by side, and the decrease of the illumination of each image will be compensated for the most part by light spread from surrounding images.

Thus, in summary, it has been shown that, indeed, for a small light source, the optics of the eye "focuses" it at the posterior pole with resulting irradiance being a function of eccentricity. On the other hand, in a diffuse light environment, retinal irradiance appears to be relatively homogeneous. The significance of this light distribution information is that if we look at a light source such as the sun, it is more focused in the macula than further in the periphery and thus more likely to cause damage. Since, in general, we do not look at point sources of light, but instead are exposed to rather diffuse lighting conditions, the studies on diffuse lighting are a more realistic counterpart of our normal environmental lighting conditions. We thus must look for other explanations for the susceptibility of the macula to disease other than the potential of increased irradiance from focused light.

Retinal Blood Flow

The retina is supplied by two circulations: the choroid, providing for the outer retina, and the retinal vasculature, supplying the inner retina. Although these two circulations are both derived from the ophthalmic artery, they have quite different characteristics. The choroid is controlled by the autonomic nervous system and has "leaky" capillaries, whereas the retinal circulation has no autonomic nervous control, shows autoregulation of blood flow, and its capillaries have tight junctional complexes in their endothelial cells. In the cat, choroidal blood flow is approximately 20 times greater than retinal blood flow.[18] We recently measured retinal blood flow in humans and found it to be in the range of 75 μL/minute.[19] As a consequence of the very high rate of blood flow through the choroid, oxygen extraction is very low. This results in the choroid providing an extremely high oxygen gradient to the outer retina. It has been hypothesized that the high choroidal blood flow is necessary to provide for the extremely high metabolic needs of the photoreceptor-RPE complex and also to provide for thermal homeostasis in the retina.

Intraocular Oxygen Gradients

Oxygen is increasingly implicated as a factor in light damage. Short wavelength visible light damage is thought to be mediated by an oxygen-dependent, photosensitized oxidation. For this reason it

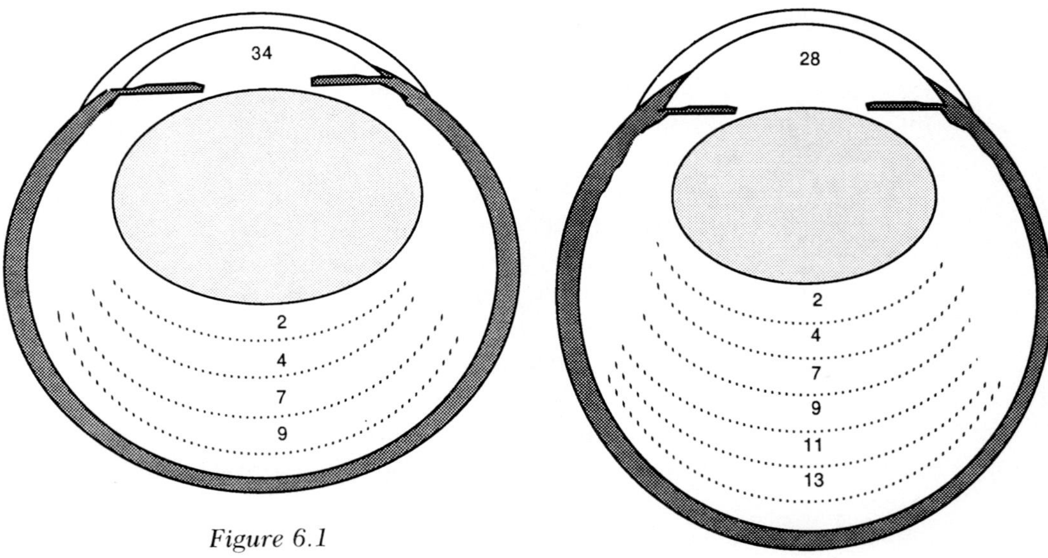

Figure 6.1

Figure 6.2

Figure 6.1. Oxygen tension distribution in the rabbit's eye.[20] Dotted lines represent oxygen tension isobars. The diagram is based on a histologic section of a rabbit eye, and the various parts of the eye are drawn to scale in relation to each other. All oxygen measurements in Figures 6.1 to 6.4 were made with oxygen microelectrodes through a pars plana approach in anesthetized animals.

Figure 6.2. Oxygen tension distribution in the cat eye.[20] Dotted lines represent oxygen tension isobars. The diagram is based on a histologic section of a cat eye, and the various parts of the eye are drawn to scale in relation to each other.

is important to know the oxygen microenvironment of the intraocular tissues. Oxygen tension gradients throughout the rabbit and cat eye have been measured using oxygen-sensitive microelectrodes.[20] The value of Po_2 was slightly higher in the anterior chamber of the rabbit (34 mm Hg) than the cat (28 mm Hg) (Figs. 6.1, 6.2). This was interpreted as possibly reflecting the greater reactivity of the rabbit anterior segment vascular system upon inserting the oxygen microelectrode into the anterior chamber. The values for Po_2 in the vitreous showed a gradient from the avascular rabbit retina (12 mm Hg) and vascularized cat retina (15–25 mm Hg) to a low of 2 to 3 mm Hg behind the lens. The oxygen gradient across the avascular rabbit retina decreased smoothly from the choroid to the inner retina (Fig. 6.3), whereas in the cat retina there was a peak in the inner retina (Fig. 6.4), reflecting the inner retinal vasculature. These findings can be interpreted as being indicative of oxygen being transported to the retina and

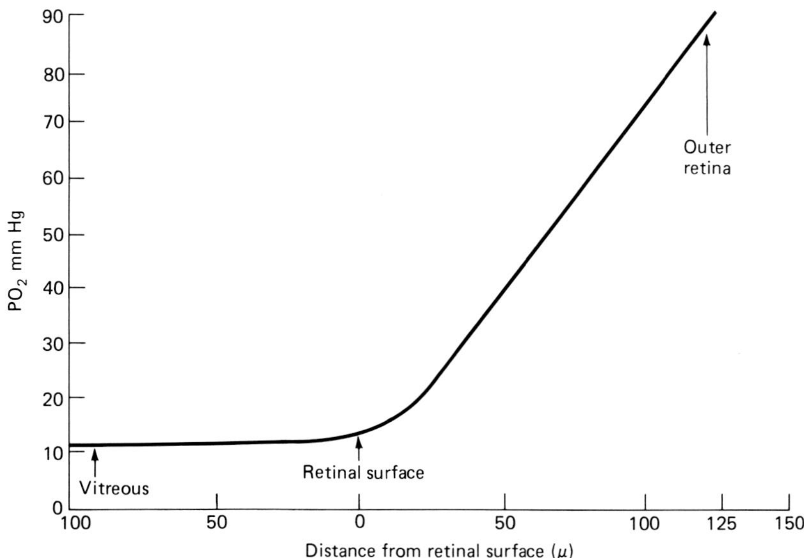

Figure 6.3. Oxygen tension profile measured across the rabbit retina.[20] Note that the oxygen tension is not uniform in the retina, but ranges from a high of Po_2 = 80 mm Hg in the outer retina, to a low of Po_2 = 12 mm Hg in the nonvascularized inner retina. This is a representative profile and shows the averaged values for a single penetration and withdrawal. Measurements were made in the light.

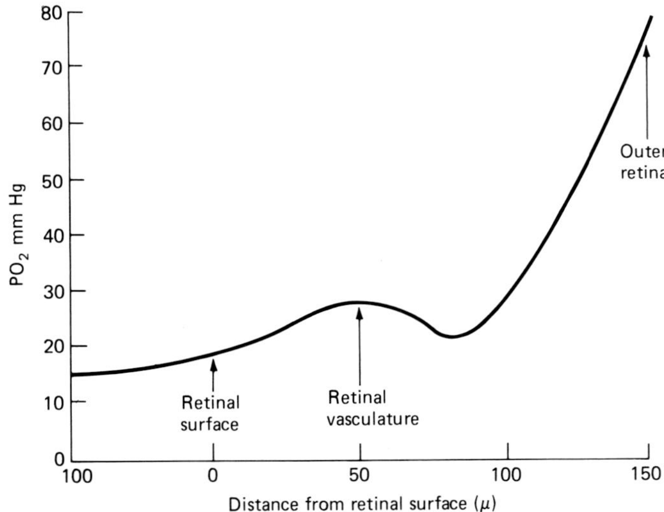

Figure 6.4. Oxygen tension profile measured across the cat retina.[20] Note that the oxygen tensions are not uniform in the retina. The inner retinal Po_2 increase is secondary to the retinal vasculature, which was lacking in the retina of the rabbit (see Figure 6.3). This is a representative profile and shows the averaged values for a single penetration and withdrawal. In these studies, oxygen tension varied from a high of Po_2 = 75 to 80 mm Hg at the level of the RPE to a low of Po_2 = 15 to 20 mm Hg at the level of the outer plexiform layer. The inner retinal Po_2 varied from 15 to 70 mm Hg, depending on proximity to retinal arterioles. These measurements were made in the light.

vitreous from the choroid, with an additional quantity being supplied by the retinal vasculature in the cat. Furthermore, Figures 6.1 and 6.2 suggest that oxygen from the ciliary processes probably supplies the metabolically active lens epithelium, but not the posterior aspect of the lens. The aqueous humor flux from the posterior chamber to the anterior chamber probably overcomes any tendency for the oxygen from the ciliary processes to diffuse into the vitreous. In the rabbit, which has an essentially avascular retina (excepting the vascularized visual streak) and therefore a lower inner retinal Po_2 than the vascularized cat retina, the lens extends closer to the retina posteriorly than in cats, monkeys, or humans (species with vascularized retinas). One could speculate that it is necessary for the posterior aspect of the lens to be closer to its oxygen source in those species lacking an inner retinal vasculature. Using the same line of speculative reasoning, one could hypothesize that a decrease in the choroidal oxygen (or other nutrients) would result in degeneration of the posterior subcapsular lens fibers. Perhaps this is the mechanism of posterior subcapsular cataract formation in diseases with choroidal degeneration such as retinitis pigmentosa and high myopia.

The transretinal oxygen profile is not uniform. The values for the rabbit retina vary from 75 to 85 mm Hg in the outer retina to 12 mm Hg in the nonvascularized inner retina (Fig. 6.3). The cat retina has a bimodal Po_2 profile (Fig. 6.4) reflecting the retinal vasculature. The values for the cat retina vary from 75 to 85 mm Hg in the outer retina to 15 to 25 mm Hg in the inner retina. In regions of the inner retina near retinal arterioles, the Po_2 is higher and may approach 65 to 70 mm Hg. To further characterize the retinal Po_2 profiles shown in Figures 6.3 and 6.4, retinal metabolism was studied. Since the early studies by Warburg,[21] the retina has been recognized as having the highest rate of oxygen consumption in the body compared with other tissues. Under normal conditions the underlying physical principles that determine oxygen concentration throughout the retina are diffusion of oxygen from the choroidal and retinal vasculatures and its subsequent utilization by retinal tissue. The high choroidal blood flow, with its low arteriovenous oxygen difference, serves as an oxygen source for the outer retina and could influence the inner retina, which has a vasculature capable of autoregulating in response to changes in oxygen concentration. We have measured oxygen consumption in the various retinal layers and have found that the photoreceptor-retinal pigment epithelial complex accounts for two thirds of total retinal oxygen consumption.[22,23] This high oxygen consumption was shown to be necessary to maintain the high levels of active sodium transport at the photoreceptor inner segment and the concomitant interstitial dark current.[22] Furthermore, this high photo-receptor oxygen consumption is decreased by light, leading to a situation in which there may be wide variations of oxygen diffusing into the inner retina.[24]

The photoreceptor-RPE complex is thus exposed to an extremely high oxygen concentration, which places it at risk for oxidative damage. This appears to be a hazard related to the variable high oxygen needs of the photoreceptors. The inner retina, on the other hand, appears not to have such wide swings in oxygen consumption, and under normal circumstances it is maintained in a much lower, relatively constant oxygen environment. Autoregulation of the inner retinal circulation helps to maintain a constant inner retinal oxygen level despite fluxes of oxygen diffusing from the outer retina.[25]

Protection

Repair Mechanisms and Lipofuscin

Biological renewal is one of the more important mechanisms that all organisms utilize for protection against the "wear and tear" of living. We are all familiar with the protective aspects of the constant turnover of the skin epithelium—the only other tissue, besides the eye, that is constantly exposed to light. Young nicely summarized this important repair mechanism as it relates to the eye.[26] Most cells of the retina undergo renewal of their molecular components on a regular basis. Cellular organelles can be turned over by means of autophagy (the mechanism by which a cell ingests cellular components and digests them using lysosomal enzymes). The photoreceptor outer segments are renewed in a unique manner. On a daily basis a certain fraction of the outer segment is shed and phagocytosed by the RPE (Fig. 6.5). The need for the rapid outer segment turnover (approximately 10–15 days) is probably related to the high degree of photo-oxidative damage they endure. Thus the RPE cell must not only renew all of its own constituents, but also carry the burden of digesting the photoreceptor outer segments. This results in the accumulation of lipofuscin granules in the RPE (Fig. 6.6).

Lipofuscin represents tertiary lysosomes or residual bodies, the indigestible end product of phagocytosis and autophagy. Lipofuscin increases with age in the RPE in a sigmoidal pattern[27,28] (Fig. 6.7). A rapid buildup of lipofuscin is noted in the first 20 years of life, followed by a relative plateau until approximately age 50, at which time a further increase is noted. We hypothesize that

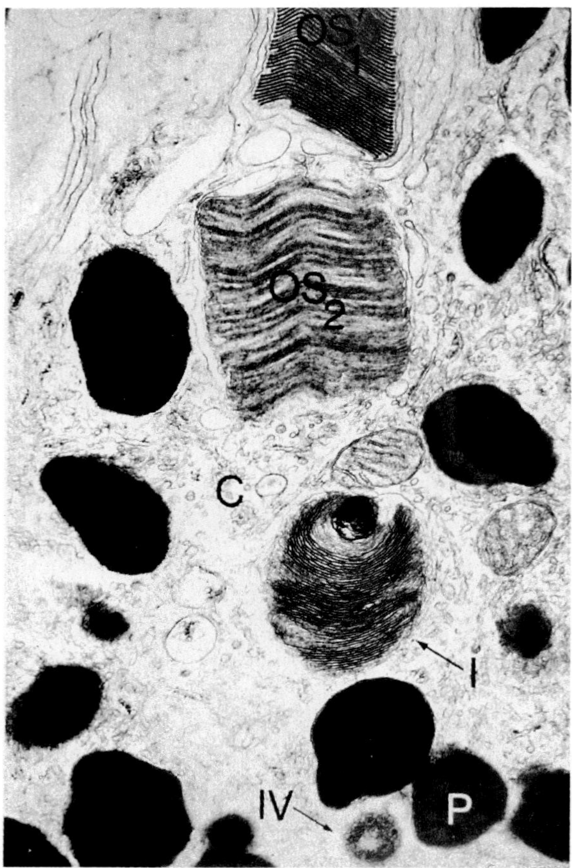

Figure 6.5. Stages of photoreceptor phagocytosis by the RPE. OS_1 is the photoreceptor outer segment. OS_2 is a recently formed outer segment phagosome recently incorporated into the RPE cytoplasm (C). (I) is a secondary lysosome showing the early stage of digestion of the outer segment discs. (IV) is a tertiary lysosome showing the end stage of digestion (lipofuscin) ($\times$ 28,000). (From Hogan MJ, Alvarado JA, Weddell JE: Histology of the Human Eye. WB Saunders Co, Philadelphia, 1971, p 422).

the rapid lipofuscin buildup in the first two decades of life is related to increased exposure of the retina to shorter UV wavelengths (295–400 nm), since the crystalline lens develops a yellow pigment that absorbs the shorter wavelengths on a time scale that matches the RPE lipofuscin buildup.[29] Interestingly, RPE lipofuscin concentration is not homogeneous in its topographical distribution, but increases from the equator to the posterior pole with a consistent dip in the RPE lipofuscin content in the foveomacular region (Fig. 6.8).

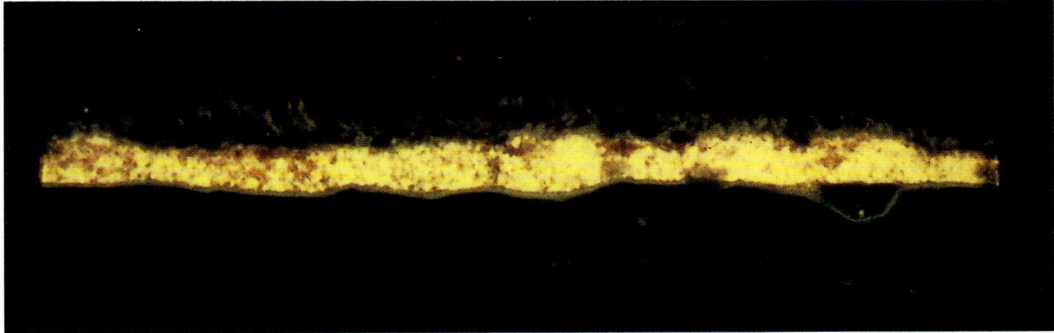

Figure 6.6. Fluorescence photomicrograph showing the autofluorescence of RPE lipofuscin. The neurosensory retina is to the top and the choroid to the bottom. (Unstained 10 μm section; × 400.)

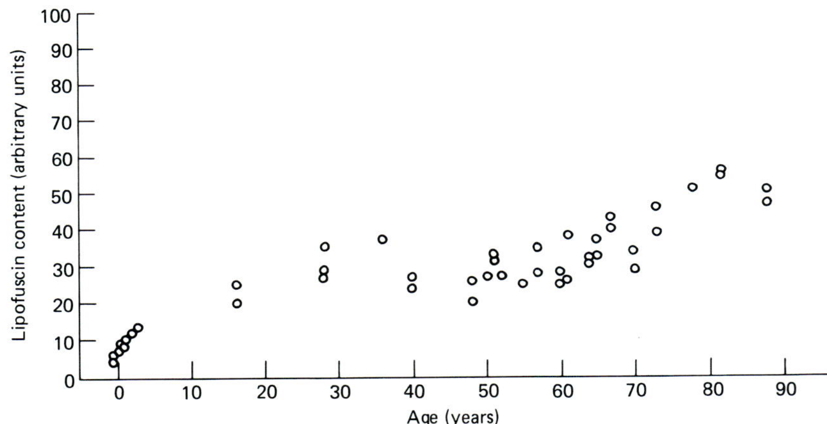

Figure 6.7. Lipofuscin content in the RPE posterior pole plotted as a function of age. From Wing et al.[27]

The question of whether lipofuscin accumulation is toxic to the RPE has not been decisively answered, although inferential evidence strongly suggests that it is.[30] If lipofuscin is formed within lysosomes after phagic sequestration of damaged or aged membranes and organelles and represents undigested residue, then lipofuscin is analagous to the accumulation of undigestable residues in lysosomal storage diseases, since in both cases the appropriate catabolic enzymes are missing for complete degradation. Feeney-Burns has shown that the percentage of cytoplasmic space decreases with increasing lipofuscin.[31] We have shown that RPE cell height increases with age and is correlated with the increases in RPE lipofuscin[28] (Fig. 6.9). Similar findings have been noted by

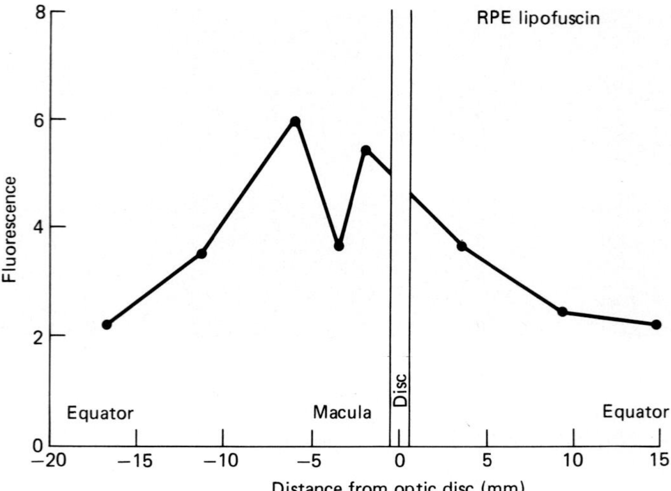

Figure 6.8. Topographical distribution of lipofuscin in the RPE. Note the buildup of lipofuscin concentration from the equator of the eye toward the posterior pole with a dip in concentration at the macula. From Weiter et al.[28]

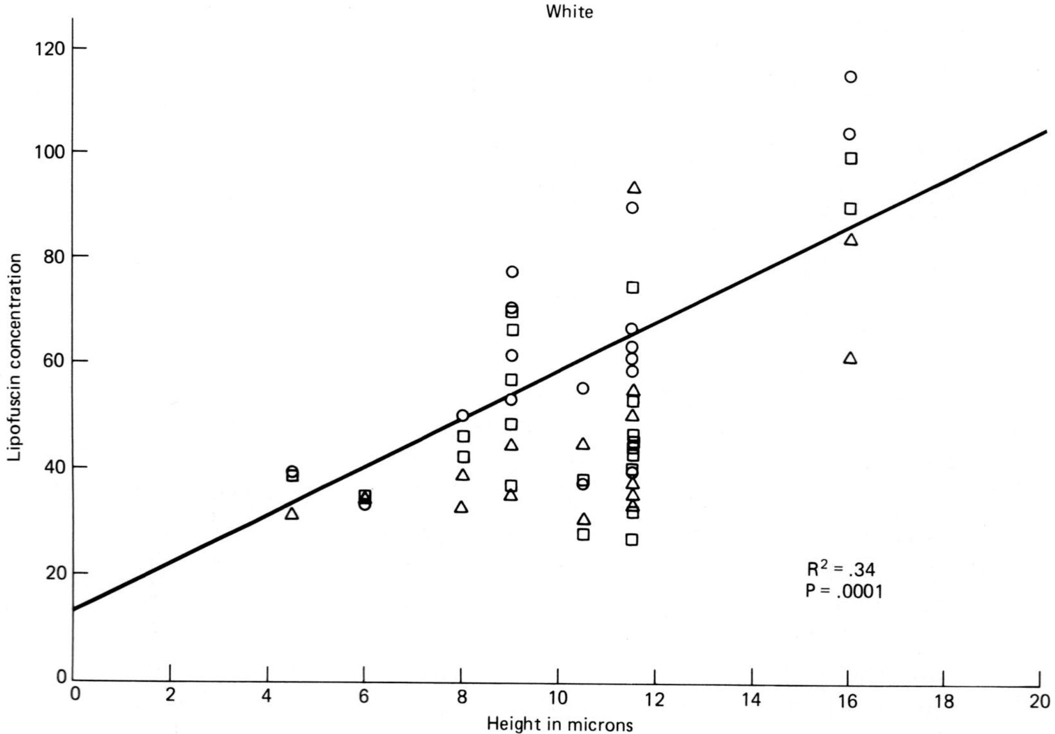

Figure 6.9. Lipofuscin concentration in RPE cells as a function of RPE cell height.

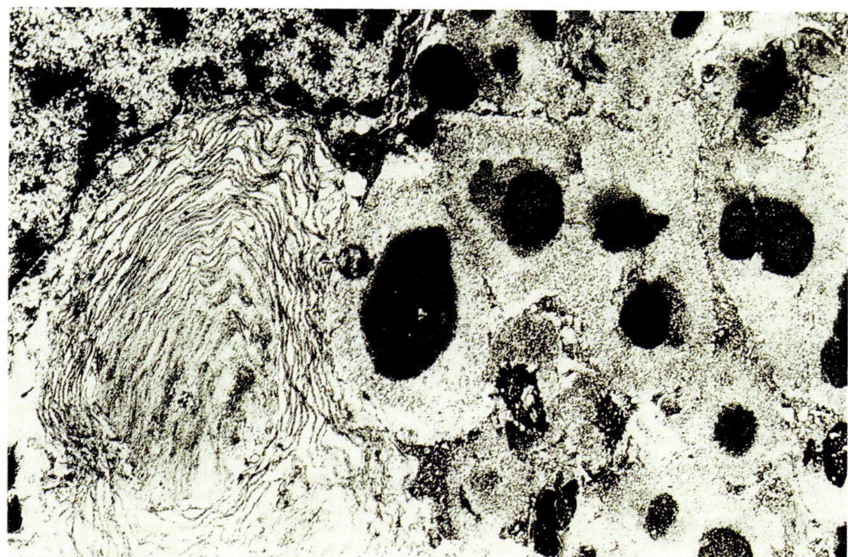

Figure 6.10. A poorly digested phagosome is noted beneath the nucleus of the RPE cell of an older individual, near the basal cell membrane in a cell with a high lipofuscin content. Undigested phagosomes are found in the apical aspect of RPE cells of younger individuals; the common finding of poorly digested phagosomes in the basal aspect of RPE cells with high lipofuscin content is suggestive of cellular dysfunction. (Electron micrograph × 30,000)

others in the rat RPE.[32] We hypothesize that just as in lysosomal storage diseases, the accumulation of undigestible residue (lipofuscin) leads to compromise of cellular function (Fig. 6.10). Supporting evidence comes from the examination of diseased eyes in which lipofuscin accumulation is excessive. Figure 6.11 shows the topographical lipofuscin profile in an eye with senile macular degeneration (SMD). The lipofuscin concentration at any RPE site was several times higher than normal for this age and was higher than ever found in a normal eye. The RPE cells at the posterior pole had already degenerated (Fig. 6.12). The remaining RPE cells were swollen and contained diminished RPE melanin. It is very tempting to speculate that these RPE cells became so engorged with lipofuscin that they "ruptured" or degenerated from cytoplasmic crowding.

The histology of a similar case, also noted to have marked elevation in RPE lipofuscin has been described.[33] Another eye with SMD, but with intact RPE in the posterior pole, was also found to have increased lipofuscin but not as marked as in Figure 6.9. Eyes from two further individuals who died in their third and fourth decades from neoplastic disease were also studied. Both

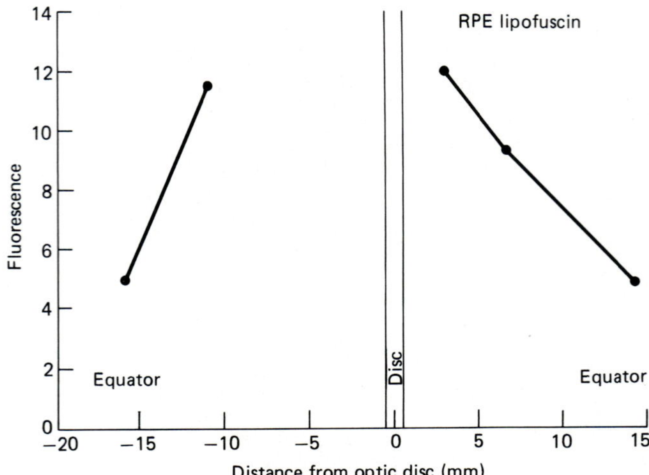

Figure 6.11. Topographical distribution of RPE lipofuscin concentration in a 70-year-old white man with SMD and posterior pole atrophy. Note that lipofuscin fluorescence is absent centrally because of atrophy of the RPE. Compare the RPE lipofuscin concentration in this diseased eye with lipofuscin concentration of a normal eye as shown in Figure 6.8. From Weiter et al.[28]

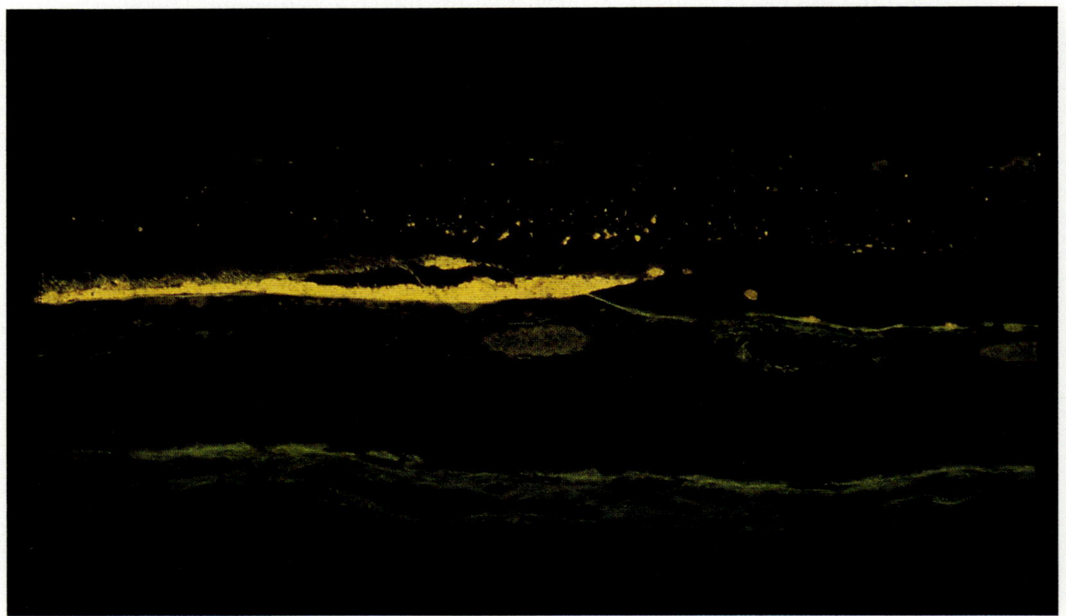

Figure 6.12. Fluorescence photomicrograph showing the autofluorescent RPE lipofuscin in the eye depicted in Figure 6.11. Retina is to the top and choroid is at the bottom. Note the swollen RPE cells choked with lipofuscin to the left and at the border of the area of RPE atrophy on the right. (Unstained 10 μm section; × 200.)

individuals received extensive antimetabolic therapy, and in both cases the RPE lipofuscin concentration was approximately twice that of normal.[28] Thus, there appears to be increasing evidence that lipofuscin buildup in the RPE appears to reflect conditions of retinal damage. The degeneration of RPE cells with exremely high levels of lipofuscin suggests that at a certain level of lipofuscin buildup it contribures to cellular dysfunction.

Melanin

The eye contains melanin derived from both neural ectoderm (retinal pigment epithelium, ciliary epithelium, and iridic epithelium) and neural crest (choroid and iris stroma). The RPE melanosomes are synthesized in utero and remain virtually unchanged thereafter. Rarely, premelanosomes are found in normal retina, suggesting a possible slow turnover of melanosomes in mature RPE, although not to a significant degree.[34] Mature melanosomes appear to be degraded by incorporation within lysosomes to form melanolysosomes and melanolipofuscin.[31] This process is slow, however, and incomplete because of the inherent stability of the melanin granules. Little is known about uveal melanin, but most likely it is similar to RPE melanin in its limited synthesis capabilities after development.

The distribution of melanin within the RPE and choroid is shown in Figure 6.13.[28] In the RPE, melanin decreases from the equator to the posterior pole, with an increase in the macular region. Choroidal melanin, on the other hand, shows an increase in density from the equator to the posterior pole. Although there is a large variability of melanin from cell to cell in the RPE and from individual to individual, on average the RPE melanin content was the same in both blacks and whites[28] (Table 6.1). The difference was in the choroid, where blacks have, on average, approximately twice the amount of melanin as whites. This makes sense, since, as noted above, the melanin in these two tissues is derived from two distinct regions embryologically, the neural crest and neuroepithelial cells. The neurocrest is the origin for melanocytes that migrate and provide pigment for such sites as skin, hair, and uvea. Such pigmentation shows marked racial variability. Since the neu-

Table 6.1. Macular pigments.

	Whites (18)	Blacks (24)
RPE Lipofuscin	573 ± 369	348 ± 92
RPE Melanin	0.29 ± 0.11	0.29 ± 0.11
Choroidal melanin	0.77 ± 0.49	1.44 ± 0.75

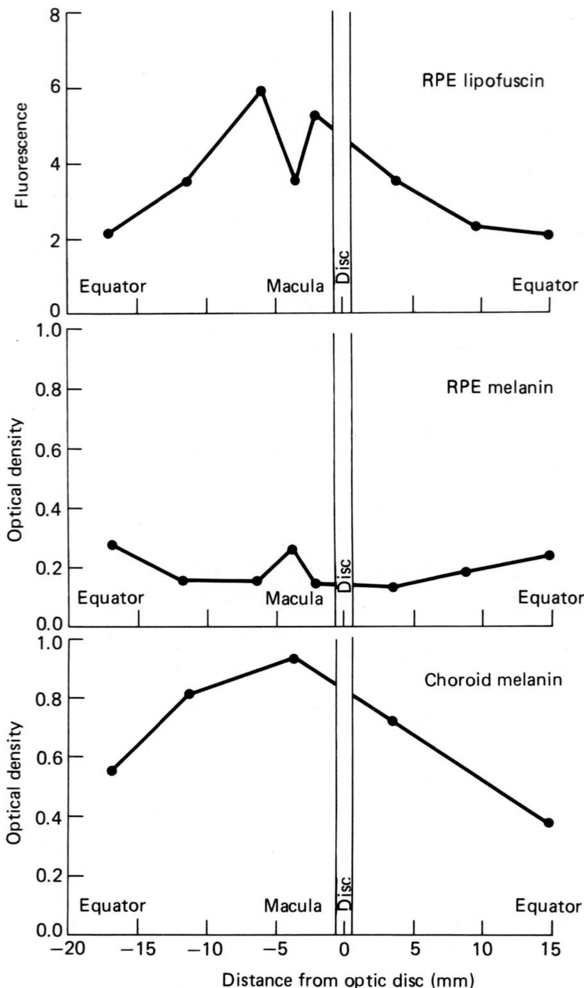

Figure 6.13. Topographical distribution of pigments of the RPE and choroid in a 60-year-old white man. This topographical distribution is representative of all the eyes studied. Total lipofuscin content (in arbitrary units) and total melanin density of RPE and choroid are displayed as a function of distance from the optic disk. Note the inverse relationship between lipofuscin and melanin content in the RPE. From Weiter et al.[28]

roepithelium provides melanin for pigmented structures of the central nervous system such as the ganglion substantia nigra, it is not surprising that the RPE, which is derived from neuroepithelium, does not show racial differences in pigmentation. The RPE melanin tends to decrease with age in both whites and blacks. This decrease was most pronounced in whites after age 50. Between individuals, fundus pigmentation closely corresponds to iris

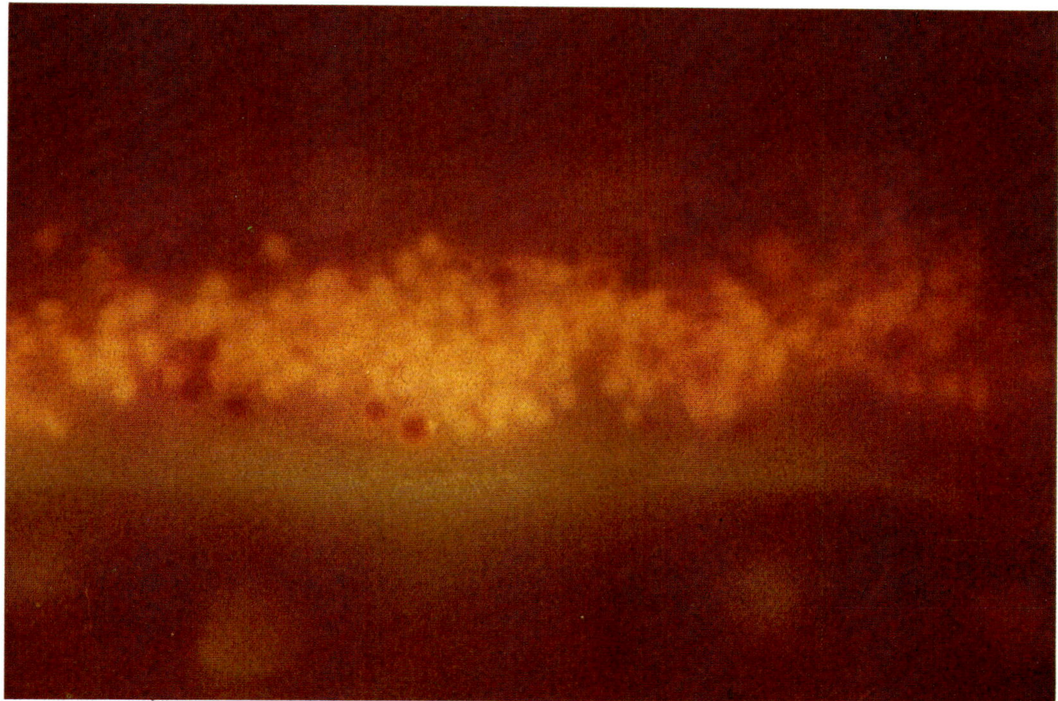

Figure 6.14. Fluorescence photomicrograph from a 60-year-old white individual. Note that there are no free melanosomes, rather the melanin is incorporated into melanolipofuscin granules. This is common after age 50. Retina to the top, choroid to the bottom. (Unstained 10 μm section; × 2,000.)

pigmentation[35] a finding not unexpected since both tissues are part of the uvea.

The exact biologic function of melanin has not yet been clarified. One role that does seem to be universally recognized, although not completely understood, is that melanin acts as a filter in protecting against light damage. When albino and pigmented rats were compared in terms of their susceptibility to retinal light damage under similar light conditions, pigmented rats showed considerable resistance to light damage. These studies suggest that the inherent susceptibility of the retina to light damage is the same between the two species and that ocular pigmentation protects against damage primarily by lowering retinal irradiance.

Other important protective properties of melanin are being understood today. These include scavaging free radicals and excited molecules, electron transfer, and affinity for drugs and metals.[36] The possibility of melanin to act as a biological electron-exchange polymer, by means of its capacity for oxidation and reduction, should protect a melanin-containing cell against reducing or oxidizing conditions that might otherwise set free, within living cells, reactive free radicals capable of disrupting metabolism.[37] This type of metabolic function ascribed to melanin (to serve as a free radical trap) is in part analogous to the protective role played by beta carotenes in photosynthetic systems, where they may act as quenchers of excited-state species of oxygen.[38] Increasingly, studies are showing that the protective action of melanin against light is not entirely a result of the absorption of light, but is related to its oxidation-reduction properties. These studies[39] further suggest that melanin granules may change their functional properties if the granules become coated with proteins. The decreased protective effect by coated melanosomes is important because in human RPE, by the age of 50, most of the melanin is incorporated into melanolipofuscin grandules[28] (Fig. 6.14). This change in the human RPE melanosomes could contribute to the relationship of SMD to ocular pigmentation.[35] Notwithstanding the mechanism of protection, it is becoming apparent that ocular pigmentation plays a role in protecting against retinal light damage.[28,35,37]

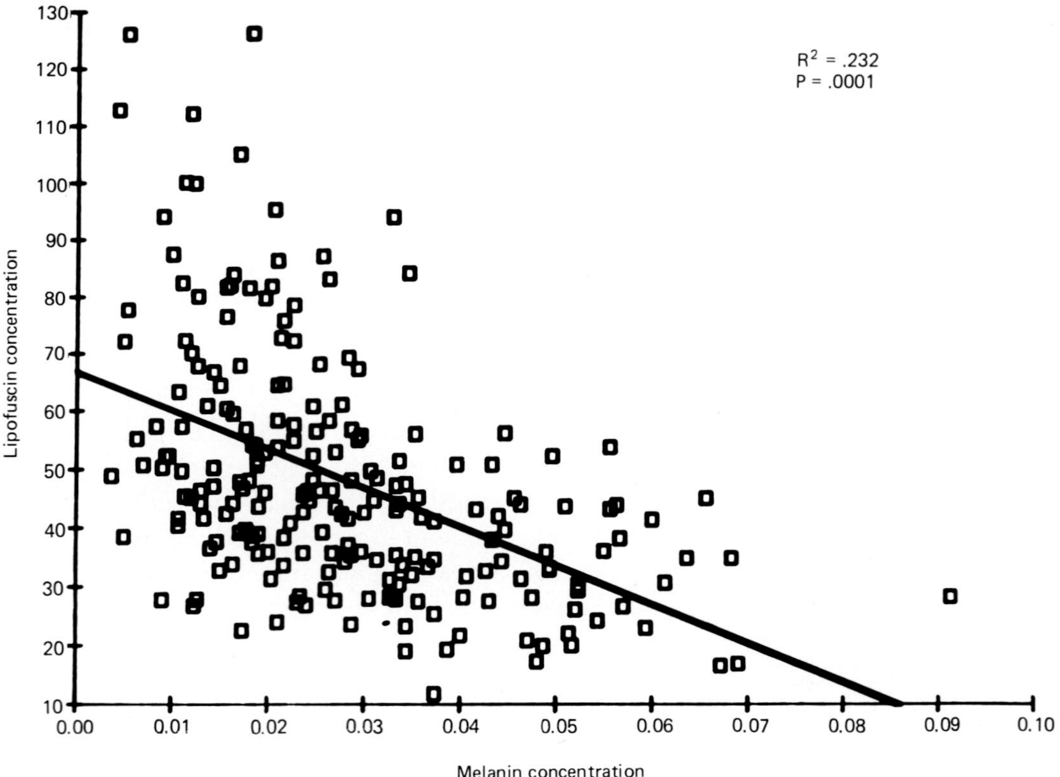

Figure 6.15. RPE lipofuscin concentration expressed as a function of RPE melanin concentration, measured apically or basally at the same sites. Included are both blacks and whites. From Weiter et al.[28]

Relationship Between RPE Melanin and Lipofuscin

Light damage mediated through photosensitized oxidations has been postulated to result in lipid peroxides and probably in the accumulation of RPE lipofuscin.[40] Since melanin protects against light damage, there could be a relation between melanin concentration and lipofuscin accumulation in the RPE. Optical measurements of the pigments of RPE and choroid were made on human autopsy eyes of both blacks and whites, varying in age between 2 weeks and 90 years old.[28] Lipofuscin can be visualized in RPE cells as yellow granules when viewed with fluorescent microscopy (Fig. 6.6). There was an inverse relationship between RPE lipofuscin concentration and RPE melanin concentration at the cellular level (Fig. 6.15) and topographically across the eye (Fig. 6.13).

The inverse relationship between RPE melanin and lipofuscin is intriguing and may suggest a protective mechanism in the formation of lipofuscin. Lipofuscin formation is a complex process most likely involving lipid peroxidation. Many factors such as light, oxygen, and nutrients probably play a role. Melanin could provide photoprotection not only by direct absorption of light, but also by serving as a scavenger of light-induced free radicals.[41] It is unlikely, though, that this scavenger effect is the sole mechanism, since whites have a much greater RPE lipofuscin content despite the fact that both whites and blacks have an equal content of RPE melanin (Table 6.1).

Light damage would be related to the total accumulated light exposure during one's life. The difference in RPE lipofuscin between whites and blacks is probably related to the differences in their choroidal pigmentation. A photon of light that was not absorbed by the photoreceptor-RPE complex would have a greater possibility of being reflected by the deeper layers of the fundus, and thus having a second pass through the photoreceptors in whites rather than in blacks. Indeed, fundus reflectance (white light) in whites is approximately 5% versus 1% in blacks.[42] This would be increased by the additional contributions from intraocular light scatter and light penetration through the iris and sclera. Whether this variation in light exposure can account for all the differences in RPE lipofuscin between whites and blacks is an unanswered question.

Macular Yellow Pigment

The central region of the human retinal posterior pole contains a yellowish area known as the macula lutea (Fig. 6.16). This yellow color is due to a carotenoid that can be found throughout the

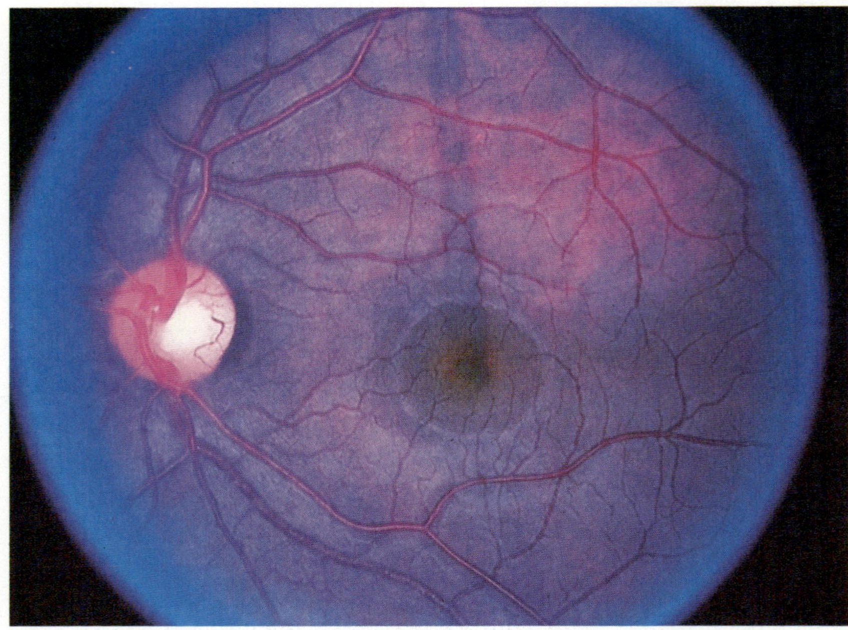

Figure 6.16. Fundus photograph of a 25-year-old individual. Note the macular yellow pigment centered on the fovea of approximately 0.5 disk diameter.

retina but has its highest concentration in the foveal region. Wald[43] identified this yellow pigment as luetein, a xanthophyll originating in the green leaves of plants. All carotenoids, including macular yellow pigment, found in man are derived from the diet. When monkeys are maintained on a xanthophyll-free diet from birth, no macular yellow pigment can be detected, in contrast to monkeys fed a standard diet including xanthophyll.[44] The macular pigment has a maximum absorbance in the blue wavelengths (460 nm) and is most dense in the photoreceptor axon layer and inner plexiform layer, with density declining markedly with retinal eccentricity[45] (Fig. 6.17).

What are the functions of macular yellow pigment? The most widely accepted role of this pigment is to reduce chromatic aberration and therefore improve visual acuity.[46] Another important role may be protection of the retina from light damage, since it absorbs wavelengths that are known to be especially damaging to the retina. This is consistent with studies of chronic light damage in monkeys showing that the area of greatest damage is perimacular with sparing of about 2 degrees in the central macula.[47] Interestingly, there appears not to be biologic variation in humans in the macular yellow pigment density based on light exposure. Studies of West Indian and European populations correlating the den-

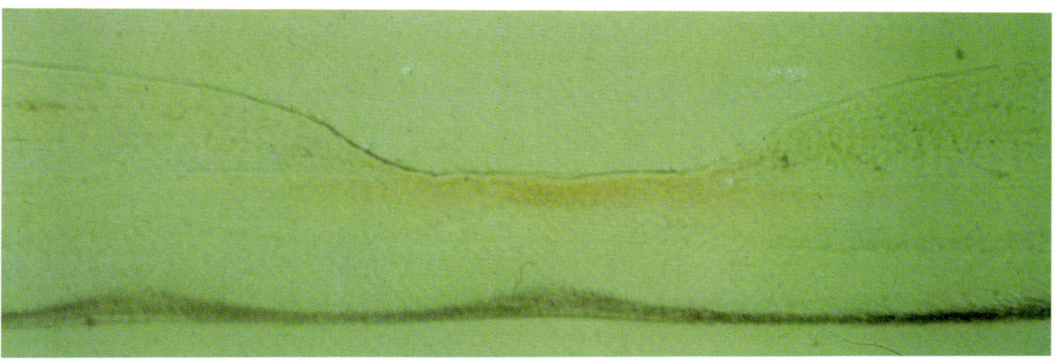

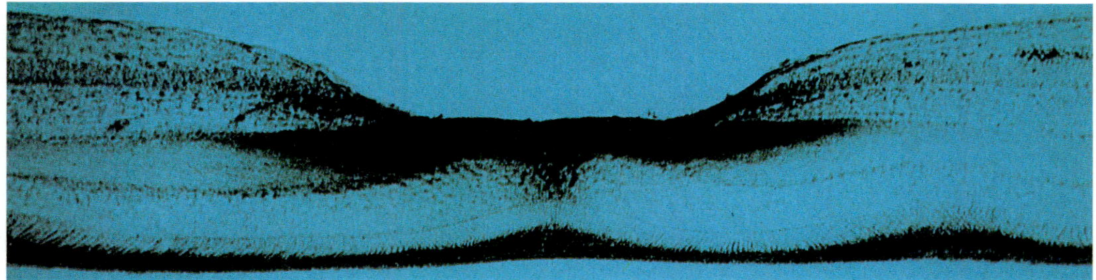

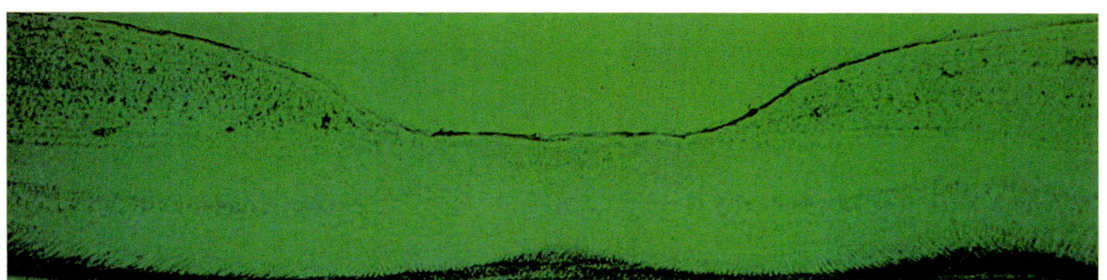

Figure 6.17. The fovea of an adult Macaca mulatta monkey. *Top:* Color photograph with white light shows the yellow color and distribution of macular yellow in the fovea. *Middle:* Color photograph with blue illumination. Pigment is highly absorptive and appears dark. *Bottom:* Color photograph with green illumination shows minimal absorption by yellow pigment. (Unstained 50 μm section; × 40.) (Courtesy of Drs. Max Snodderly and Francois Delori.)

sity of macular pigment with ethnic group, environment, age, and color of skin, hair, and eyes showed no difference except for a higher density in red-haired individuals.[48]

Antioxidant Protection in the Retina

The human retina has features that make it susceptible to photo-oxidation. The photoreceptor membranes, containing high levels of polyunsaturated fatty acids, exist in a milieau of extremely high oxygen tension (Figs. 6.2 and 6.3) and are exposed to light. In spite of these features contributing to oxidative stress, the human retina is capable of good function over 80 or more years. This fact bespeaks of the protective mechanism discussed above in the human retina. Additional protection is afforded by antioxidants. Several antioxidant mechanisms have been characterized in the retina, including vitamin E, glutathione, selenium-dependent glutathione peroxidase, non-selenium-dependent glutathione peroxidase, catalase, and superoxide dismutase. Probable antioxidant roles for ascorbate (vitamin C), carotenoids and melanin are also present.

Studies of deficiencies of these antioxidants play an important role in bringing attention to oxidative stress and retinal damage. Previously, it has been shown that vitamin E deficiency caused central retinal degeneration in the monkey.[49] Interestingly, some of these dietary effects can be noted in nature. The photoreceptors from cattle fed dry rations, as occurs in the summer in California, had undetectable levels of vitamin E and were highly susceptible to oxidative damage.[50] In contrast, abundant vitamin E was found in the photoreceptors from cattle being fed green rations, as occurs in the winter in California, and these latter photoreceptors were resistant to oxidative stress. This finding illustrated that vitamin E content of the photoreceptors might be susceptible to variations in dietary vitamin E intake.

In the rat a deficiency of both vitamin E and selenium results in a marked lipofuscin buildup in the RPE, but not other retinal tissues.[51] Lipofuscin is thought to be indigestible residues of peroxidized lipids and therefore indicative of oxidative stress. Since lipofuscin accumulation is also noted to occur in human RPE during aging (Figs. 6.6 and 6.7), we felt it would be instructive to look at the topographical buildup of lipofuscin in vitamin E-deficient rats. Similar to the aging human RPE, the vitamin E-deficient rat shows an increase in lipofuscin in the RPE, with the greatest increase at the posterior pole (Weiter JJ and Dratz EA, unpublished data) (Fig. 6.18). Note, however, that the lipofuscin buildup at the posterior pole does not show a dip such as is found in humans at the foveal-macular region. This most likely reflects the fact that the rat does not have a macula and the protective macular yellow pigment (see bull's eye maculopathy).

In spite of increasing experimental work linking oxidative stress to retinal damage, there are limited studies looking at the role of oxidative stress in human retinas. Since antioxidant deficiencies are rare in humans, and since disease states such as SMD, where oxidative stress is thought to play a major role, take approximately

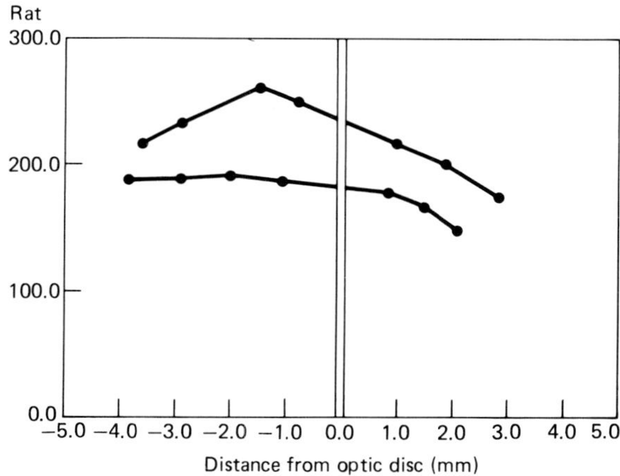

Figure 6.18. Topographical distribution of lipofuscin in the RPE of the rat. Top curve from a vitamin E-deficient animal; bottom curve from control animal. Note similarity of top curve to that of human RPE (Figure 6.8) except for the foveal dip, explained by the absence of a fovea in the rat. (In collaboration with Dr. Edward Dratz.)

60 years to occur, the linkage between oxidative stress and retinal damage in humans will be more difficult to prove. We attempted to correlate various dietary factors such as vitamin E and selenium with SMD. Our studies on vitamin E were nonconclusive and showed marked variations in plasma vitamin E levels among patients with SMD. We attributed this to the widespread use of vitamin E supplements by today's elderly. There is a suggestive role, though, for selenium nutrition in SMD. We have found that the degree of SMD severity is inversely correlated with the plasma level of selenium-dependent glutathione peroxidase.[52] Selenium-dependent glutathione peroxidase is a selenium-dependent enzyme, and its activity is an indicator of the adequacy of the selenium nutritional status. These early, preliminary studies are highly suggestive of a role for oxidative stress in causing retinal damage in humans.

Disease States

Solar Retinopathy

Solar retinopathy, or eclipse blindness, is caused by gazing at the sun and has been a recognized entity for many years. The clinical entity known as foveomacular retinitis, which tends to occur in young military personnel, is also felt to represent solar retinopathy. In solar retinopathy the patient usually notes immediate symptoms of central scotoma, chromatopsia, and metamorphopsia. The vision is usually reduced to the range of 20/40 to 20/80. The immediate clinical findings are usually minimal, consisting of either no obvious changes or subtle edema and occasionally a small hemorrhage. At about 24 hours a characteristic foveolar lesion is present, consisting of a small yellow-white spot in the fovea (Fig. 6.19a). Fluorescein angiography is typically normal, although occasionally an area of hyperfluorescence can be seen (Fig. 6.19b). With time, this lesion fades and what remains is a smaller lamellar hole or cyst in the region of the fovea (Fig. 6.20). Most likely, the lesions are initially at the level of the RPE and of the macular yellow pigment, with secondary involvement of the other retinal elements. Histological studies of eyes after several hours of sun gazing showed varying degrees of damage to the RPE, including irregular pigmentation, necrotic cells, and edema, but the inner and outer segments of the photoreceptor cells appeared normal.[53] Approximately 50% of patients with solar retinopathy recover 20/20 vision at 6 months after exposure.

Recent research provides convincing evidence that the cause of solar retinitis is closely related to the photochemical effects of the short wavelengths in the solar spectrum rather than to thermal effects from an indiscriminate mixture of both visible and infrared radiation.[54] This is based on experimental measurements of retinal temperatures showing that the maximum temperature induced in the retina by sun gazing does not exceed a few degrees Celsius.[55] It should be noted, however, that a temperature rise of 3° to 4°C above ambient in the retina may produce thermally enhanced photochemical damage. The solar spectrum at sea level peaks at 470 nm.[54] The solar energy is greater between 400 and 500 nm than in any subsequent 100 nm bandwidth. Since photochemical lesions tend to be caused by short wavelengths, sunlight is an extremely important cause of photochemical lesions.

Senile Macular Degeneration

Background

The retina is susceptible to photo-oxidative damage because of its local environment, i.e., high O_2 gradients, chronic light exposure, and the nature of the lipid membranes, which are susceptible to photo-oxidation.

However, the retina has a number of protective mechanisms. For example, under conditions of high light levels, one might think that corresponding lower O_2 levels could be protective. In-

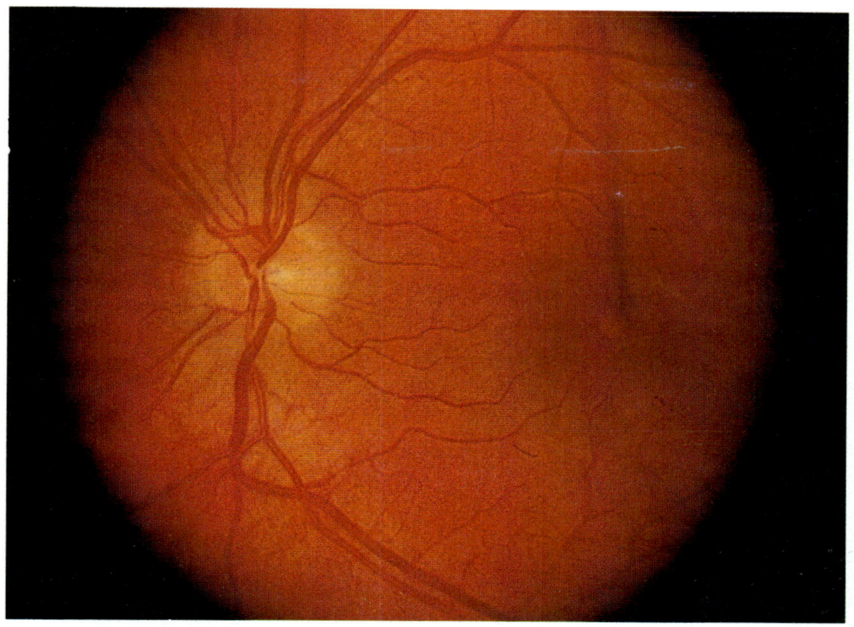

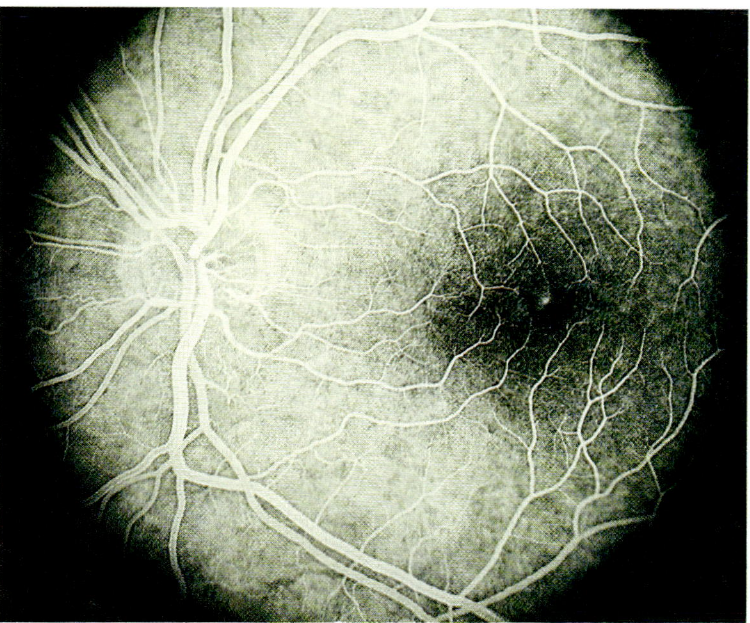

Figure 6.19. **A.** Color fundus photograph of the left eye of a 16-year-old girl who 48 hours earlier had gazed at the sun for over an hour. Note the characteristic small yellowish spot in the fovea. Clinically, this yellowish spot appeared to be in the neurosensory retina, probably representing absorption by the macular yellow pigment. Vision was 20/60. **B.** Fluorescein angiogram of the same individual showing a smaller hyperfluorescent spot corresponding to the clinical yellowish spot.

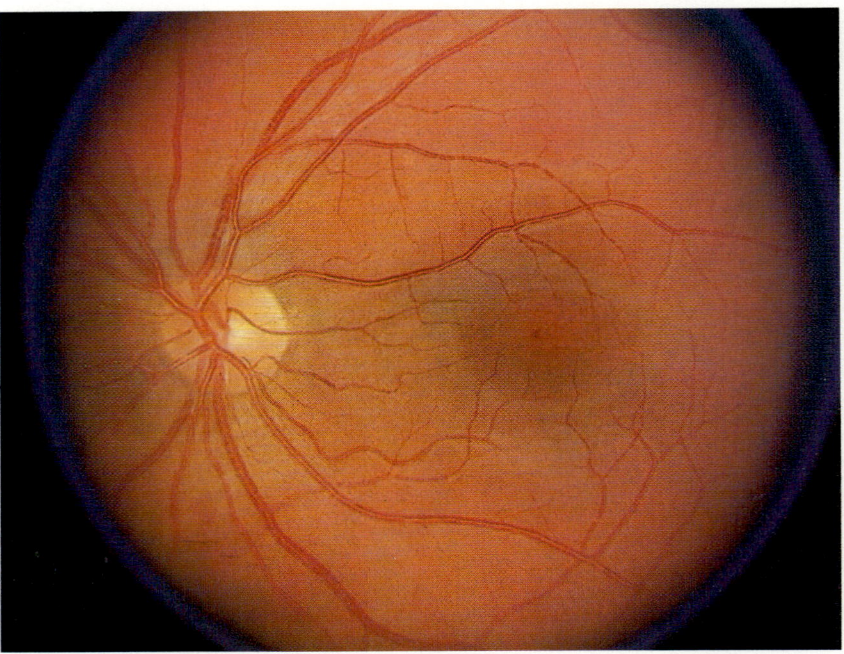

Figure 6.20. Color fundus photograph of a 19-year-old individual who had documented solar retinopathy 1 year earlier. Vision has returned from 20/50 to 20/20. Note that all that remains is a small foveal cyst. The fluorescein angiogram was normal.

deed, in the eye one finds that except for very active intraocular metabolic tissue (i.e., corneal endothelium, lens epithelium, ciliary epithelium, and the photoreceptor-RPE complex), the O_2 levels are quite low (the posterior lens surface, the entire vitreous body, and the inner retina) (Figs. 6.1–6.4).

Actual control of harmful light levels and wavelengths are modulated by filters within the eye. For example, the cornea holds back wavelengths lower than 300 nm. The crystalline lens blocks wavelengths below 400 nm, except in the juvenile lens, which does pass a high percentage of light between 300 to 400 nm. Thus, we can see that very little UV radiation gets to the retina. Within the retina itself, a yellow pigment, present mainly in the macula, blocks some of the blue light (i.e., peak absorbance at 460 nm). The distribution of this yellow pigment parallels the cone distribution in the retina. The melanin in the retinal pigment epithelium can be thought of as absorbing the light that has already passed the photoreceptors. This back scattered or retroreflected light would otherwise degrade the visual image or do further harm to the receptor.

For a moment, let us look at this melanin in more detail. The melanin in the retinal pigment epithelium is basically the same in composition and amount in all races. However, the amount of melanin in the choroid varies among the races. Blacks, for example, have about twice the amount of melanin in their choroid as whites (Table 6.1).

What of the photoreceptors themselves? We have seen that they are surrounded by protection devices, but do they have an intrinsic protective mechanism? Of all the elements of the retina, the outer layers of the photoreceptors are the only ones that are subject to turnover. Approximately 10% of the outer segment is renewed daily. This essentially means that the complete outer segment is replaced every ten days. Interestingly enough, this rate of turnover is the same as skin epithelium (the only other major tissue exposed to light). Figure 6.5 illustrates this turnover of the outer segment very graphically.[56]

The photoreceptors and the retinal pigment epithelium also possess an abundance of free radical scavengers such as vitamin E, superoxide dismutase, catalases, ascorbic acid, and glutathione peroxidase. Vitamin E, which is fat soluble, has one of the highest concentrations in the body in the membranes of the outer segments of the photoreceptors. On the other hand, water-soluble ascorbic acid is present is high concentrations in the cytoplasm of the retinal pigment epithelium. Finally, just as photo-oxidative spoilage affects fruit more rapidly at room temperature than at refrigerated temperature, so too photo-oxidative retinal damage is temperature dependent. Thus, there is a mechanism to carry away the heat generated when melanin absorbs visible light. In fact, the choroidal circulation, which has the highest blood flow in the body, is anatomically designed for rapid removal of excess heat.

Aging

It has been shown that the amount of protective melanin in the retinal pigment epithelium significantly decreases at about age 50.[28] This loss of melanin is paralleled with an increase in the concentration of lipofuscin (Fig. 6.7). Lipofuscin appears to be a measure of damaged retinal membranes. In fact, it represents undigested phagosomes. Thus, one might think of this series of events as the decrease in protective melanin accompanied by an increase in retinal damage as seen in the elevated lipofuscin. Ultimately, the buildup of this waste material produces cellular dysfunction. One sees a somewhat analogous mechanism of dysfunction in the lipid storage diseases. There, an enzyme for lipid membrane breakdown is absent or diminished, resulting in a buildup of intracellular waste products, which ultimately interferes with normal cell function. Turning to lipofuscin buildup, we see

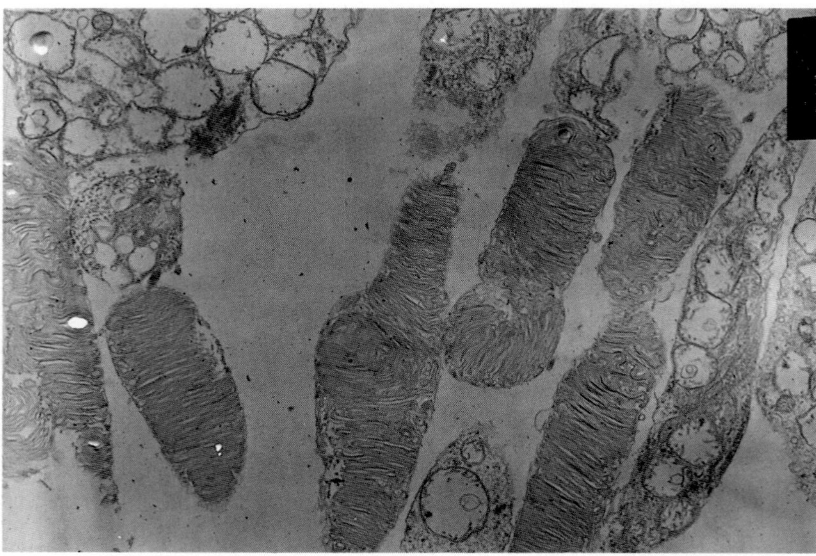

Figure 6.21. Rod photoreceptor outer segments from a 70-year-old individual. Note the irregularities in the stacking of disks, with whorl patterns and convolution of the disk stacks. These findings are common in the aging retina ($\times$ 15,000).

that it is membrane lipid converted to a form that is indigestible by normal lysosomal enzymes, thus its buildup and subsequent cell dysfunction. It should also be noted that lipofuscin buildup occurs almost twice as rapidly in whites as compared with blacks (Table 6.1). As further evidence that lipofuscin buildup is related to light damage one notes a rapid buildup in the retina of the child, where the crystalline lens allows UV light to pass through to the retina (Fig. 6.7). At this point we must admit that we have no information concerning age-related changes in the other protective mechanisms such as outer segment renewal and levels of antioxidants. It is common to find in the aging retina changes in the photoreceptors suggestive of dysfunction (Fig. 6.21). The outer segments in older eyes very frequently show irregularities in the stacking of the disks, with whorl patterns and what appears to be folding of the outer segments. The significance of these findings is yet to be determined.

Definition

The leading cause of blindness in the United States in people over the age of 65 is SMD.[57] The hallmark of this disease is the presence of exudative or soft drusen in the macula (Fig. 6.22). The spectrum of SMD includes RPE atrophy (Fig. 6.23), sub-

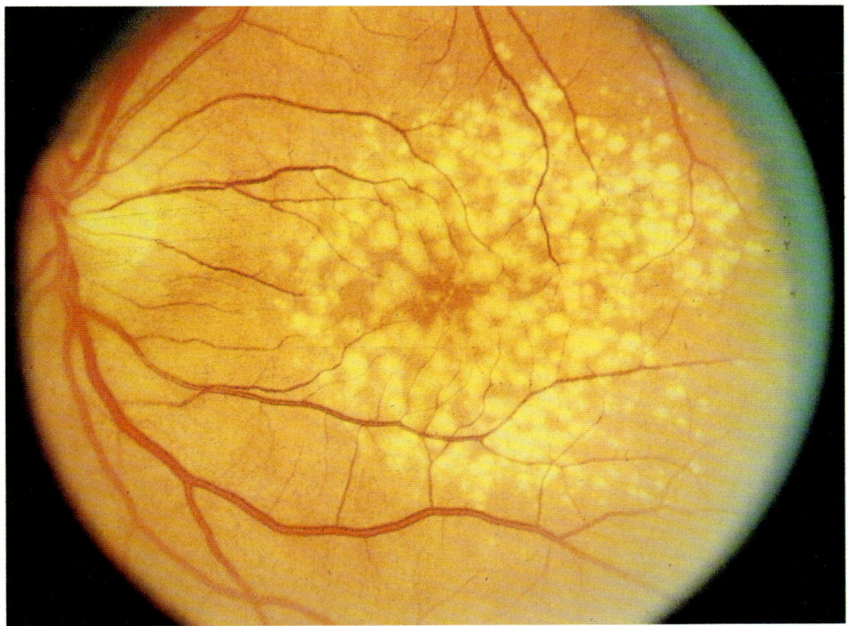

Figure 6.22. Patient with SMD associated with exudative or soft drusen. The characteristics of this type of drusen are irregular sizes and margins and a tendency toward confluencey.

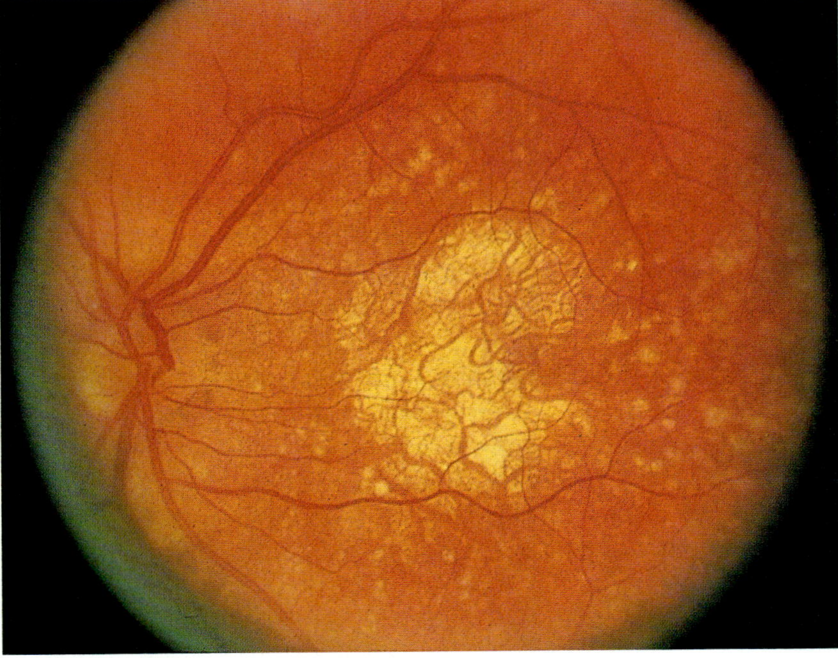

Figure 6.23. Patient with SMD associated with exudative drusen and macular atrophy.

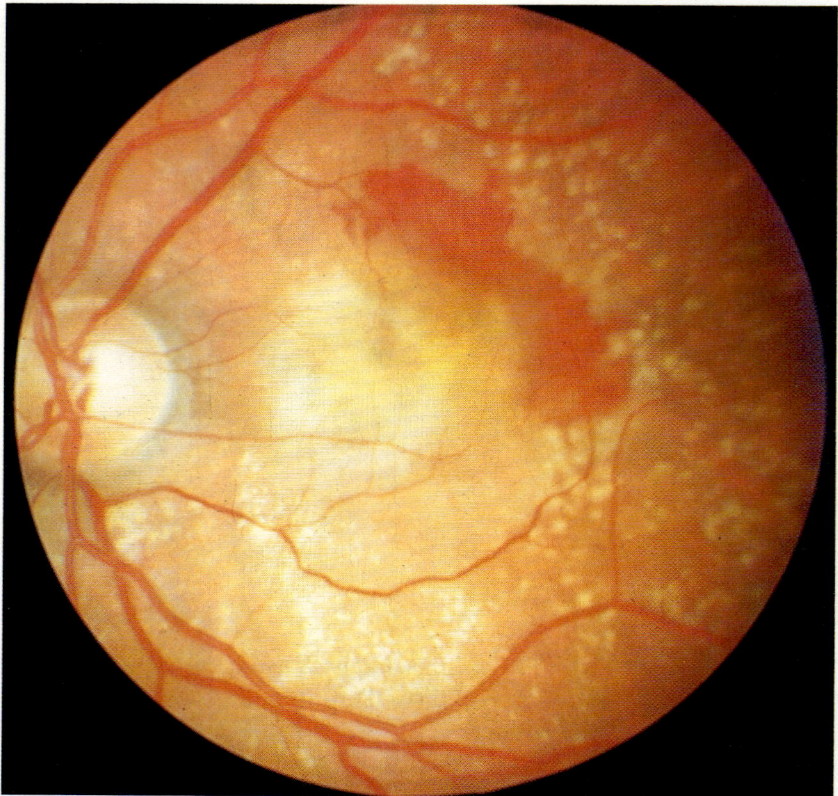

Figure 6.24. Patient with SMD associated with exudative drusen, subretinal neovascularization, hemorrhage and exudate, and a serous detachment of the retina.

RPE neovascularization and hemorrhage (Fig. 6.24), RPE detachment (Fig. 6.25), and disciform scars (Fig. 6.26).

Senile macular degeneration is a complex, age-related disease process. Its pathogenesis is poorly understood. Central to the development of SMD are drusen formation. We differentiate between cuticular or hard drusen from "exudative" or soft drusen. Hard drusen represent a nonspecific response of epithelial cells to insult,[33] and are analogous to the "guttata" of corneal endothelial cells. In contrast, exudative drusen represent cytoplasmic debris deposited in the inner aspects of Bruch's membrane (Fig. 6.27). Our histologic studies demonstrate that these are formed by extrusion of a portion of the RPE cell with its cytoplasmic contents (thus, the name exudative drusen) (Fig. 6.28). The process of apoptosis, also noted by others,[58] may be due to RPE dysfunction from excessive lipofuscin accumulation. It could represent an attempt to decrease lipofuscin concentration by exocytosis. The gradual accumulation of cytoplasmic debris results in drusen that have

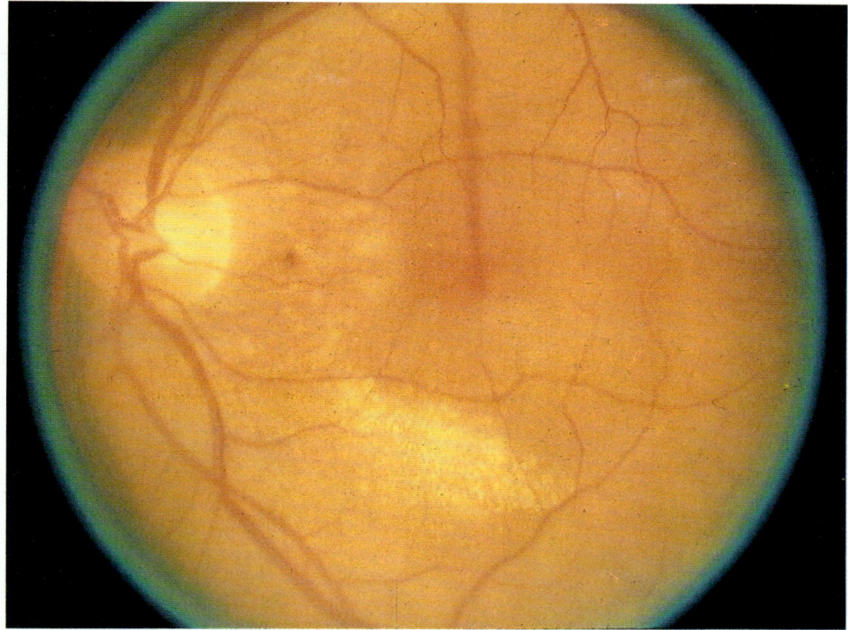

Figure 6.25. Patient with SMD associated with exudative drusen and a retinal pigment epithelial detachment. The presence of exudate and notching on the nasal margin are indicative of subretinal neovascularization.

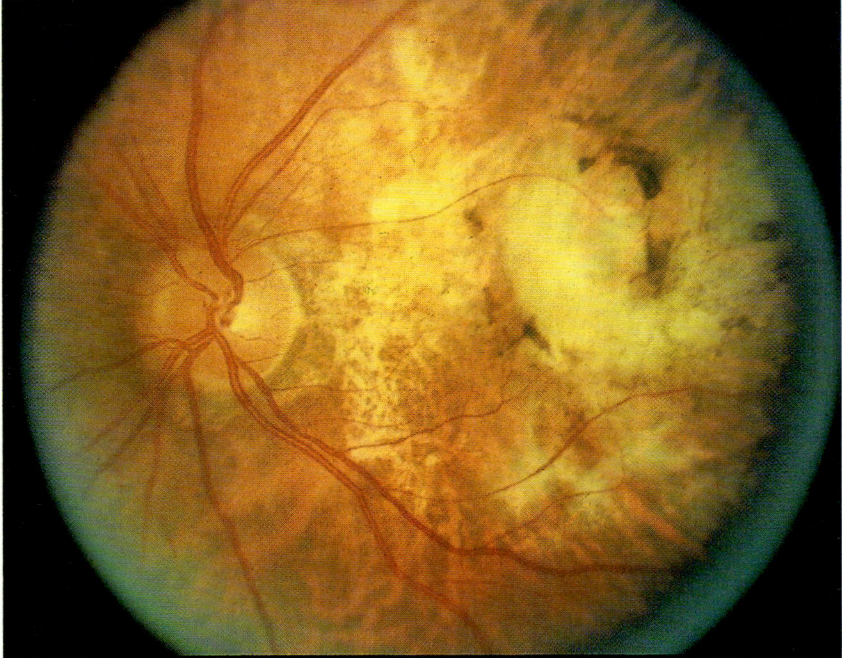

Figure 6.26. Patient with SMD associated with a fibrovascular disciform lesion and atrophy.

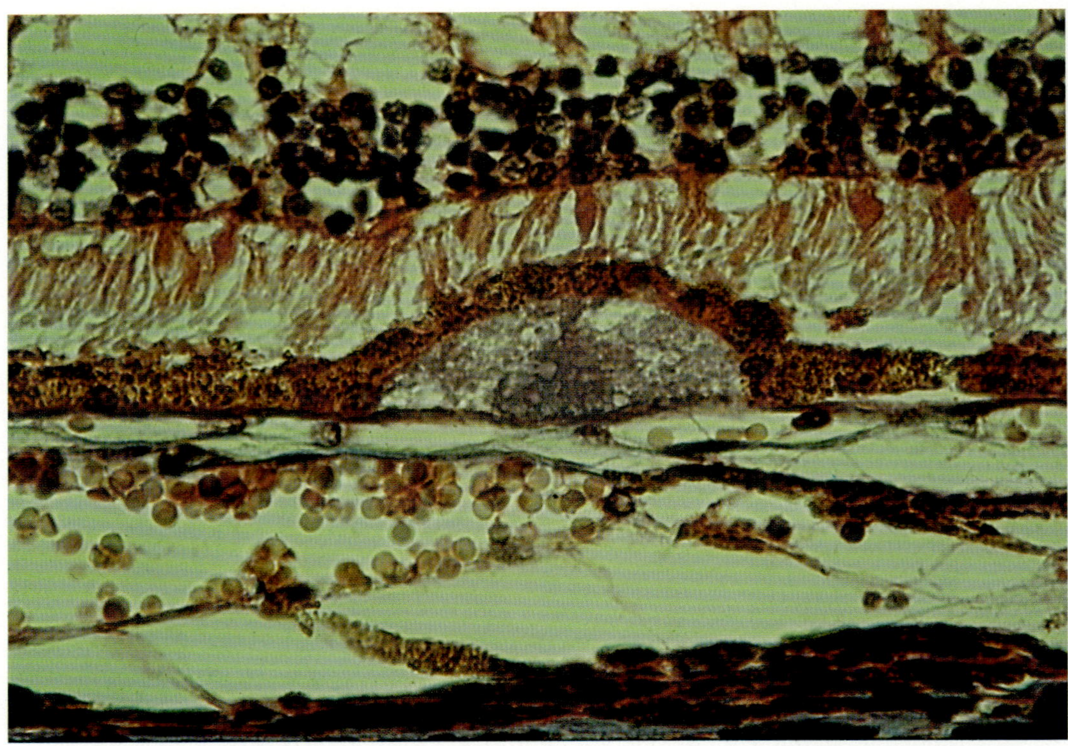

Figure 6.27. Exudative drusen showing loose-appearing vacuoles and membraneous debris deposited in the inner aspects of Bruch's membrane (10 μm section; × 400).

indistinct margins and tend to become confluent (Fig. 6.22). Ultrastructural studies show that drusen contain membrane debris, vesicles, and, occasionally, abnormal collagen (Fig. 6.29). The intermittent findings of lipofuscin, a melanolipofuscin granule, undigested photoreceptor phagosomes, and mitochondrial fragments are suggestive of extruded cytoplasm. Exudative drusen accumulate in the inner aspect of Bruch's membrane, between the basement membrane (basal lamina) of the RPE cells and the elastic layer of Bruch's membrane (Fig. 6.29). This accumulation may ultimately result in inner-layer splitting and separation of Bruch's membrane, particularly if there is associated hydrostatic pressure of leaking

---

Figure 6.29. Exudative drusen. Note the accumulation of vacuoles and ▷ membraneous debris in the inner aspect of Bruch's membrane between the basement membrane of the RPE and the calcified elastic layer of Bruch's membrane. The cytoplasm of the RPE shows marked accumulation of lipofuscin (electron micrograph × 25,000).

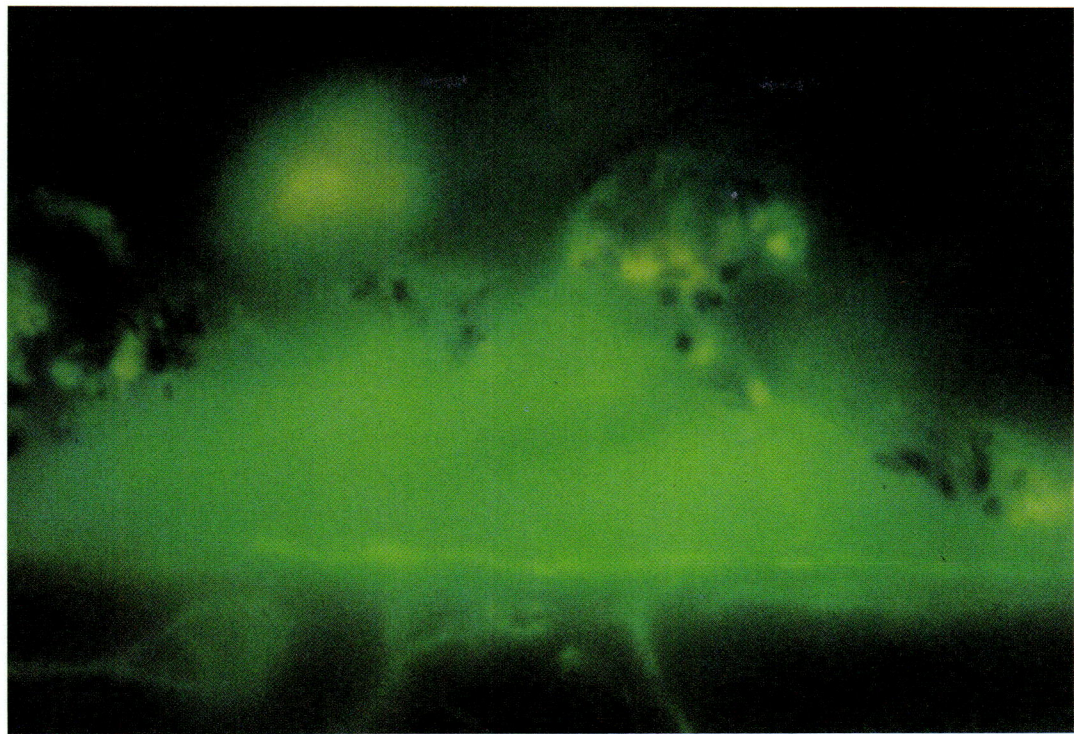

Figure 6.28. Fluorescence photomicrograph of drusen with overlying degenerating RPE. Note the extrusion of melanin and lipofuscin into the drusen. (Unstained 10 μm section; × 2,000.)

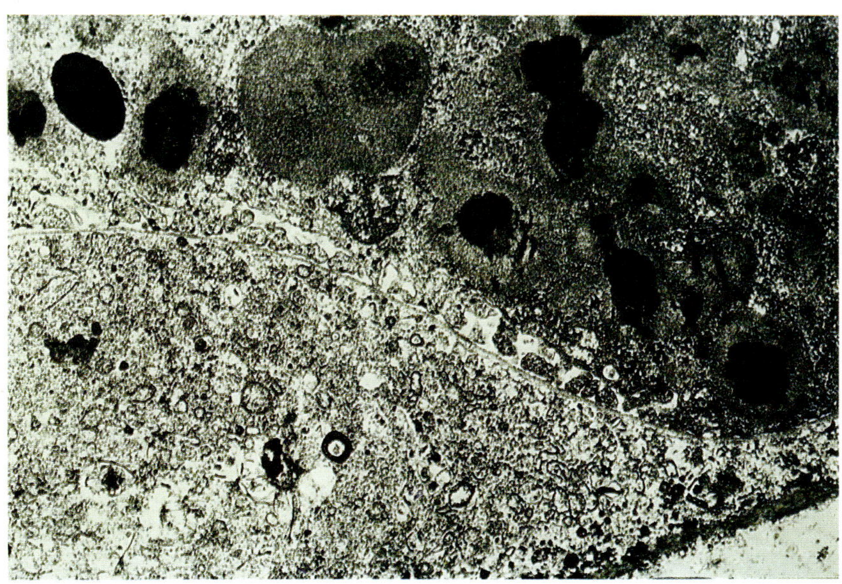

serosangenous fluid from neovascularization.[59] The result would be a RPE detachment, the pathogenesis of which has been previously discussed.

The formation of exudative drusen reflects a diffuse disorder of the RPE, and, indeed, overlying RPE cells show attenuation and atrophy. Associated with the RPE dysfunction is a concomitant atrophy of the choriocapillaries.[33] Most likely, this combination of RPE dysfunction, the imposition of a diffusional barrier secondary to drusen, and choriocapillary atrophy, serves as a stimulus for neovascularization in SMD.

Although this entity is an important cause of blindness, we do not really understand its origin. However, evidence tends to implicate chronic light damage as playing a major role in this condition. For example, many studies have shown that UV and short wavelength, visible blue light can cause retinal damage in animals. Other studies have shown that pigmentation in animals protects against both experimental and natural light damage. Unfortunately, it has been very difficult to show a connection between 60 years of light exposure and retinal damage in the human. However, there are some strong connections between the amount of human ocular pigmentation and the presence of SMD. Specifically, we have shown that SMD is rare in the pigmented races (i.e., blacks and Asians). Among whites, the presence of SMD correlates highly with the level of ocular pigmentation.[35] Statistically there is a significant increase in SMD in blue-eyed people as compared with brown-eyed people. However, although melanin pigment is protective, it does diminish with age. Therefore, one would expect that brown-eyed whites would ultimately develop SMD. In fact, our study shows that the average onset of SMD in brown-eyed whites is 5 years later than in blue-eyed individuals.

If light is a contributing factor to SMD, then the blockage of light should be protective. Thus, one might ask, would the incidence of SMD be diminished in a group of people who wore spectacles (spectacle glass blocks a portion of the UV spectrum)? In fact, it has been shown that there is a marked decrease in SMD in spectacle-wearing, myopic patients (Weiter JJ, unpublished data). Among the small group of myopes who develop SMD, it appears that these patients first started wearing spectacles after age 25. Recall that the crystalline lens transmits UV light up until age 20–25. Therefore, it would appear that blockage of UV light in the young eye may protect against SMD in the aging adult. Nature's laboratory yields further anecdotal support for this idea. For example, we have followed several black patients with unilateral aphakia secondary to trauma as children. These patients as older adults showed signs of SMD only in the aphakic eye. It should further be noted that these patients did not wear corrective spectacles after their cataract surgery.

Up until now, we have presented some evidence that the presence of pigmentation and the decrease of light play significant roles in protecting against SMD. Is there any evidence that alteration in antioxidant levels are related to SMD. A recent finding of ours shows an inverse relationship between the level of the plasma antioxidant, glutathione peroxidase (a selenium-dependent enzyme that can be altered by diet), and the severity of SMD.[52] Thus, there is increasing evidence for the role of sunlight in the pathogenesis of SMD. Although the evidence is not absolute, it is of such a degree that preventive measures should be considered.

Melanoma

The relationship between intraocular melanoma and exposure to sunlight remains controversial. In a recent case-control study,[60] individuals with intraocular malignant melanoma were compared with matched controls to evaluate the role of sunlight exposure as a risk factor for the development of this tumor. The factors that were found to correlate statistically with the occurrence of uveal melanoma were 1) having blue eyes, 2) having been born in a southern climate, 3) having spent a large amount of time outdoors, 4) sunbathing, 4) using sunlamps, and 5) rarely wearing hats, visors, or sunglasses in the sun. The study also suggested that wearing corrective lenses may exert a weak protective effect.

This study adds further evidence that sunlight contributes to the induction of neoplasms on the light-exposed areas of the body, namely the eyes and skin. The risk factors for cutaneous melanomas are similar to those listed above for uveal melanomas. These factors include a higher incidence in the white population in comparison to more darkly pigmented races, differences in incidence with latitude, corresponding to differences in sunlight exposure. The majority of skin melanomas appear to be more related to acute intermittent exposure or short exposures to high-intensity sunlight, rather than continual low-level sun exposure.[61] Although skin melanomas are most commonly found on the back, they do avoid the areas least exposed to sunlight. Skin melanomas as well as uveal melanomas are correlated with the age at exposure to increased sunlight. Exposure at an early age rather than later in life appears to be the critical factor.

The study relating sun exposure as a risk factor for uveal melanoma needs to be viewed cautiously. Not all uveal melanomas are to be found in areas of the eye having the greatest sunlight exposure. Ciliary body and peripheral choroidal melanomas would appear to be rather protected from light exposure unless there is a systemic as well as local effect from sunlight exposure. Furthermore, correlation studies have failed to find a consistent gradient of increasing incidence with decreasing latitude, as has been found in several studies on skin melanomas. This factor has been explained by noting that, in general, the more northern American populations have a higher representation of individuals with blue eyes whereas the southern populations have a greater representation of brown-eyed individuals. Thus the greater protection offered by brown eyes would be negated by the greater sunlight exposure.

Bull's Eye Maculopathy

Bull's eye maculopathy is an associated group of clinical entities manifested by a peculiar doughnut-shaped macular lesion at the level of the retinal pigment epithelium. Chloroquine retinopathy would be the best-known example of bull's eye maculopathy (Fig. 6.30). Table 6.2 lists many of the known causes of this condition. Although this list appears to describe a widely disparate group, there appears to be a basic finding that divides them into two major groupings. Those entities appear to be either related to the use of photosenitizing drugs, such as chloroquine, or to disease states characterized by a buildup of lipofuscin in the retinal pigment epithelium, such as ceroid lipofuscinosis and SMD.

At this point, we would like to present a hypothesis for this unusual maculopathy. Let us look at the typical lesion that presents itself as a central normal area surrounded by a ring of atrophy (Fig. 6.31). Is there an underlying anatomic substrate that can account for the appearance of this strange lesion? We believe there is. In Figure 6.8, we have shown that lipofuscin in the retinal pigment epithelium accumulates in the normal eye, such that there is a peak accumulation at the posterior pole with a depression in

Table 6.2. "Bull's eye" macular lesion.

Chloroquine toxicity
Congenital cone dystrophy
Ceroid lipoproteinosis
Senile macular degeneration
Hallervorden-Spatz syndrome
Benign concentric annular macular dystrophy
Sjögren-Larsson syndrome
Fucosidosis
Canthaxanthin toxicity
Retinitis pigmentosa
Stargardt's/fundus flavi

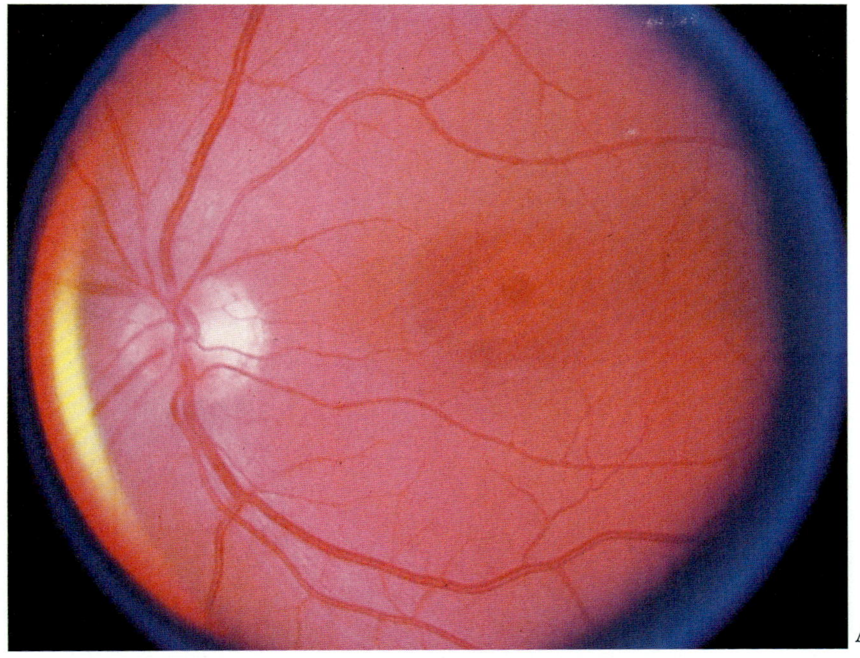

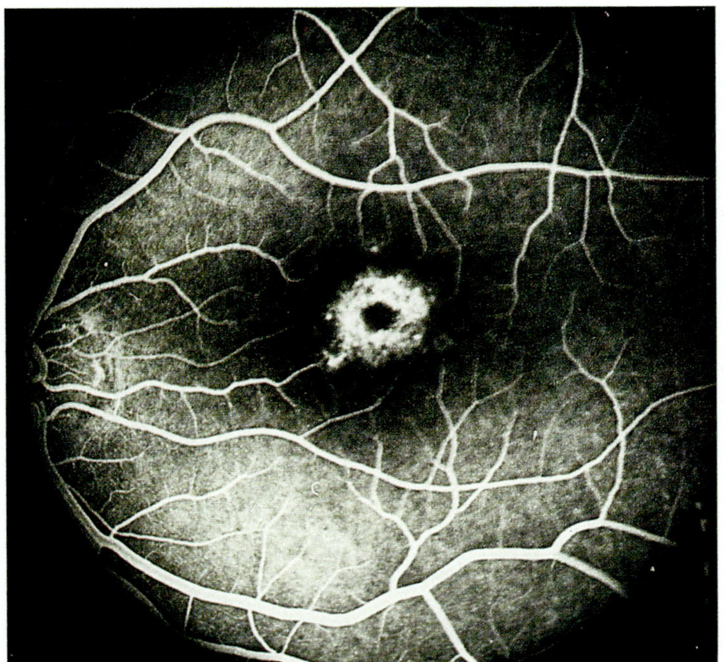

Figure 6.30. **A.** Color fundus photograph of patient who had taken chloroquine for 10 years for rheumatoid arthritis. Note the bull's eye pattern consisting of a normal foveal region surrounded by perifoveal outer retinal atrophy. Vision was 20/25. **B.** Fluorescein angiogram of same patient showing obvious bull's eye pattern secondary to perifoveal "window-defect."

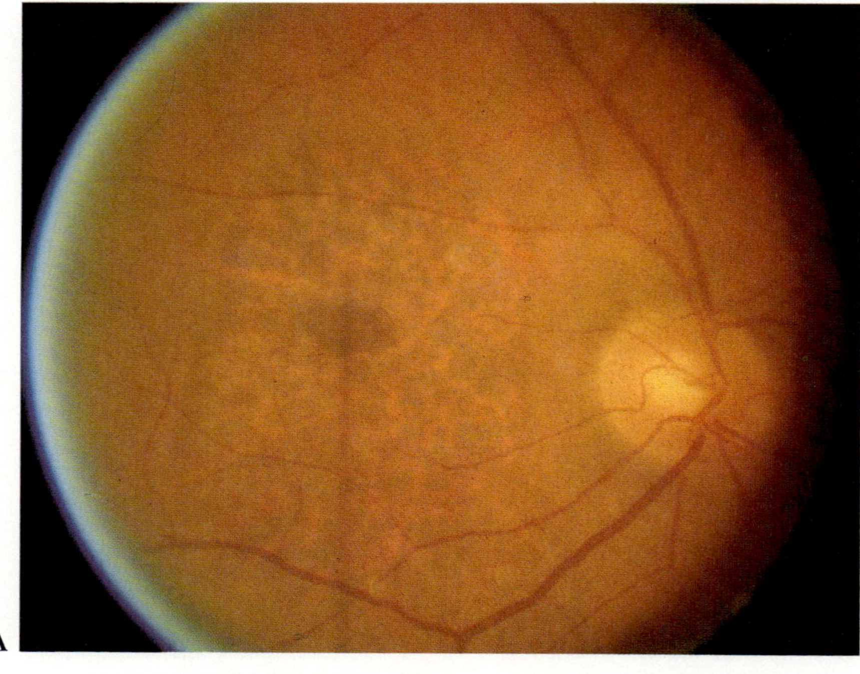

Figure 6.31. **A.** Color fundus photograph of a 79-year-old patient with SMD. Note bull's eye pattern. Vision was impaired because of large ring scotoma but measured 20/30. Note oval pattern of the bull's eye with the long axis horizontal. This is a common finding with bull's eye maculopathy and corresponds to the distribution of macular yellow pigment. **B.** Fluorescein angiogram demonstrating the extent of atrophy surrounding the bull's eye.

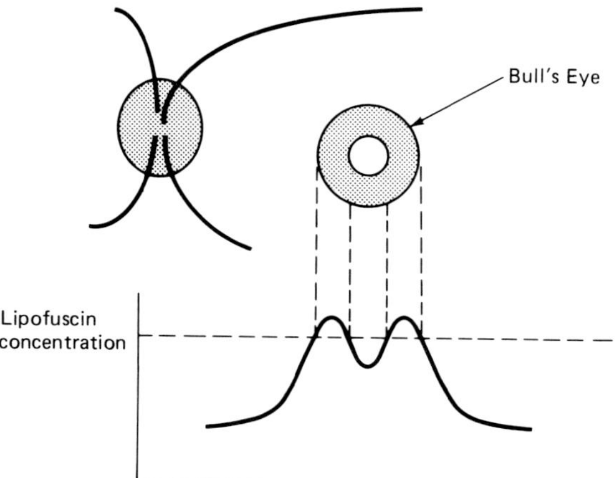

Figure 6.32. Hypothesized mechanism for the development of a bull's eye lesion. These lesions occur in conditions that result in increased RPE lipofuscin. Because of the foveal dip in the lipofuscin concentration, the surrounding regions reach excessive levels of lipofuscin concentration before the center. The photoreceptive effect of macular yellow pigment most likely contributes to this macular dip in lipofuscin concentration.

the region of the fovea. We feel that when lipofuscin reaches a sufficient concentration, it results in the degeneration of the retinal pigment epithelium. Thus, we see in Figure 6.32 that increasing the general level of lipofuscin would produce a zone of degeneration. The conditions mentioned in Table 6.1 all either work toward increasing the level of lipofuscin or are associated with a photosensitizing drug. These drugs bind to melanin and render the eye more susceptible to light damage. Thus, the increased melanin at the posterior pole sets the stage for a greater level of damage produced by the photosensitizing drugs in the same area.

At this point, one may ask how can this theory explain the small area of normal tissue in the center of the degeneration? There are two possibilities. As previously shown (Fig. 6.13), there is an inverse relationship between melanin and lipofuscin concentration in the retinal pigment epithelium. Note that as melanin peaks, lipofuscin dips at the fovea. Thus, we find protective melanin at the fovea. However, probably even more important is the presence of macular yellow in the fovea, which serves to reduce the amount of blue light striking this area.

As evidence that macular yellow plays this role, we measured the size of the zone of macular yellow in normal eyes and found that it corresponds to the size of the central area of bull's eye maculopathy. We believe that the close approximation of these values suggests that macular yellow pigment contributes to the bull's eye pattern through a photoprotective mechanism.[62]

Retrolental Fibroplasia (RLF)

Retrolental fibroplasia, or as the disease spectrum is more commonly termed today—retinopathy of prematurity—was first recognized as a clinical entity in 1942.[63] With the introduction of closed incubators in the late 1940s, the disease reached epidemic proportions and became the leading cause of childhood blindness in developed countries. Owens and Owens[64] in 1949 found a decreased incidence of retrolental fibroplasia in infants receiving vitamin E supplement, but since other investigators failed to confirm these findings, interest in the relationship between vitamin E and RLF waned. In that same year, RLF was noted to be associated with increased use of oxygen therapy, among other factors, by Kinsey and Zacharias.[65]

The oxygen hypothesis to explain the development of RLF was presented by Cambell[66] in 1951 and further supported by the controlled clinical studies of Patz et al[67] and the animal experimentation of Ashton.[68] A controlled, multicenter, national cooperative study firmly indentified oxygen as the primary etiologic factor in the development of RLF.[69]

The causal relationship between overuse of oxygen and RLF led to the rigid curtailment of oxygen in the nursery, resulting in a dramatic reduction in the incidence of RLF. Unfortunately, this decreased incidence due to curtailment of supplemental oxygen was associated with increased mortality and morbidity among premature infants, particularly those with the respiratory-distress syndrome.

The past decade has seen a dramatic increase in the incidence of RLF. This increase was due to not only the more liberal use of oxygen in the respiratory-distress syndrome to prevent mortality and brain damage, but also the lengthened survival of premature infants of very low birth weight, which was made possible by modern intensive care techniques. Thus, the risk-benefit quandry has transformed RLF from a footnote in medical history as a "new" disease that was solved by modern medical science, to a persistent, unsolved problem. Moreover, the exact role of oxygen in this disorder has been called into question since RLF can occur in infants not receiving supplemental oxygen.

Today there is renewed interest in RLF, not only because of its increased occurrence, but also because it may serve as a model for understanding both basic mechanisms of oxygen toxicity and retinal neovascular proliferative diseases. Current concepts of the pathogenesis of oxygen toxicity involve various characteristics of the developing retina that are unique to that tissue. The normal retina contains no blood vessels until the fourth month of gestation, when vascularization proceeds from the optic disk toward the periphery. Before this time, the underlying choroidal circulation provides for full-thickness retinal oxygenation.

We have hypothesized that the centrifugal maturation of the retinal photoreceptors (the major oxygen-consuming cell type of the adult retina) precedes this vascular outgrowth.[70] Their maturation and consequent increased oxygen consumption would make it impossible for the choroid to provide adequate full-thickness retinal oxygenation. This relative inner retinal hypoxia would provide an orderly controlled stimulus for the progression of retinal vascularization from the optic disk to the peripheral retina.[71] The vascular growth reaches the peripheral retina on the nasal side at approximately 36 weeks of gestation and on the temporal side at 40 weeks.[72] Hyperoxia causes arteriolar constriction and capillary endothelial-cell cytotoxicity, particularly in the immature retinal vasculature.[72]

Why are immature vessels more sensitive to oxygen toxicity than mature vessels? Oxygen toxicity can be mediated through free radical production and can be counteracted by antioxidants and free radical scavengers. Recent work has shown that when cells are gradually exposed to hyperoxia, they can build up resistance by increased intracellular production of antioxidants.[73] It is tempting to postulate that in the case of the immature retinal vasculature, the endothelial cells are at risk because they have not matured enough to produce protective enzymes or have not had the necessary gradual exposure to oxygen to induce an adequate complement of these enzymes.

The factors most related to the incidence of RLF, namely birth weight, gestational age, and duration of supplementary oxygen treatment, are all intercorrelated and reflect the functional immaturity of the retina. Recent studies[74] showing the efficacy of vitamin E in reducing the incidence and severity of RLF support the concept that this is a disease related to oxidative damage to tissue with insufficient protective mechanisms.

If this concept is true, then any mechanism that enhances oxidative damage would increase the incidence of RLF and, vice versa, anything that retards oxidative damage would lessen the incidence of RLF. The obvious additional factor to consider is the role of light. Short wavelength visible light damage mediated by an oxygen-dependent, photosensitized oxidation is well recognized. Thus it would be expected that light could enhance the oxidative damage in an immature retina.[75]

The recent report[76] showing a higher incidence of RLF in a group of infants who had been exposed to a brighter nursery lights than among infants kept under reduced lighting confirms the central role of oxidative damage in the production of RLF. Light, at levels that are normally harmless, can thus be shown to induce damage if the proper conditions exist. Thus, in the susceptible immature retina, light appears to be another factor that needs to be considered as contributing toward the incidence of RLF.

Vitreous Degeneration

In aging eyes, partial liquefaction and vitreous detachment followed by condensation of the vitreous fibrils is observed. Such vitreous changes may result in traction on the retina and lead to retinal tears, hemorrhages, epiretinal membrane formation, and retinal detachment. It has also been noted primarily from viscosity measurements that hyaluronic acid of aged and of pathologic vitreous has considerably lower molecular weight than that of young and normal vitreous.[77] Although detailed investigations are lacking, it can be reasonably implied that changes in composition, structure, and interaction properties of the virtreous components might be responsible for such abnormalities.[78] As in many other tissues, degeneration of vitreous can be caused by molecular species of active oxygen such as H_2O_2, hydroxyl radical, superoxide anion, and singlet oxygen. Nonenzymatic degradation of hyaluronic acid has been shown to occur in the presence of a reducing agent and oxygen. The reductant, ascorbic acid, is present in the vitreous in large quantities.[77] It is generally believed that depolymerization of hyaluronic acid occurs through production of free radicals during the auto-oxidation of reductants and the reduction of molecular oxygen. Recently, singlet oxygen has been shown to reduce the viscosity of hyaluronic acid as well as producing possible conformational changes in the molecule.[79] The normally low vitreous oxygen levels (Fig. 6.1 and 6.2) are probably protective against these oxygen-induced vitreous degenerative changes. In consideration of possible light damage to the vitreous, it is to be noted that the vitreous is constantly exposed to light and photodynamically generated active species of molecular oxygen. Although light below 400 nm is usually absorbed by the cornea and lens, higher wavelength light can produce active oxygen species by the sensitizers present in the normal eye. Aphakic eyes suffer more vitreous degeneration than do phakic eyes, and it has been suggested that the cause is increased passage of UV light.[77] Thus, the vitreous, similar to other ocular tissues, appears to be at risk for photo-oxidative damage. This susceptibility of the vitreous to photo-oxidative damage may help to explain the common clinical observation that rhegmatogenous retinal detachments (secondary to vitreous disease) are more common in whites than in blacks.

References

1. Duke-Elder S, MacFaul P: Radiational injuries, in Duke-Elder S (ed). System of Ophthalmology, vol XIV, London, Henry Kimpton, Chap. X, 1972.

2. Fuchs E: Text Book of Ophthalmology. Philadelphia, JB Lippincott, p 705, 1923.

3. Meyer-Schwickerath G: Koagulation der Netzhaut mit Sonnenlicht. Ber Dtsch Ophthalmol Ges 55:256–259, 1950.

4. Friedman E, Kuwabara T: The retinal pigment epithelium. IV. The damaging effects of radiant energy. Arch Ophthalmol 80:265–279, 1968.

5. Tso MOM, Fine BS, Zimmerman LE: Photic maculopathy produced by the indirect ophthalmoscope: I. Clinical and histopathologic Study. Am J Ophthalmol 73:686–699, 1972.

6. McDonald HR, Irvine AR: Light-induced maculopathy from the operating microscope in extracapsular cataract extraction and intraocular lens implantation. Ophthalmology 90:945–951, 1983.

7. Irvine AR, Wood I, Morris BW: Retinal damage from the illumination of the operating microscope; an experimental study in pseudophakic monkeys. Arch Ophthalmol 102:1358–1365, 1984.

8. Calkins JL, Hochheimer BF: Retinal light exposure from operating microscopes. Arch Ophthalmol 97:2363–2367, 1979.

9. Jampol LM, Kraff MC, Sanders DR, et al: Near-UV radiation from the operating microscope and pseudophakic cystoid macular edema. Arch Ophthalmol 103:28–30, 1985.

10. Kraff MC, Sanders DR, Jampol LM, Lieberman HL: Effect of an ultraviolet-filtering lens on cystoid macular edema. Ophthalmology 92:366–369, 1985.

11. Weiter JJ, Finch ED: Paramagnetic species in cataractous human lenses. Nature 254:536–537, 1975.

12. Weiter JJ, Subramanian S: Free radicals produced in human lenses by a biphotonic process. Invest Ophthalmol Vis Sci 17:869–873, 1978.

13. Noell WK, Walker VS, Kang BS, Berman S: Retinal damage by light in rats. Invest Ophthalmology 5:450–473, 1966.

14. Robertson DM, Feldman RB: Photic retinopathy from the operating microscope. Am J Ophthalmol 101:561–569, 1986.

15. Young RW: A theory of central retinal disease, in Sears ML (ed). New Directions in Ophthalmic Research. Yale University Press, New Haven, Conn, 1981.

16. Kooijman AC: Light distribution on the retina of a wide-angle theoretical eye. J Opt Soc Am 73:1544–1550, 1983.

17. Bedell HE, Katz LM: On the necessity of correcting peripheral target luminance for pupillary area. Am J Optom Physiol Opt 59:767–769, 1982.

18. Weiter JJ, Schachar R, Ernest JT: Control of intraocular blood flow. I. Intraocular pressure. Invest Ophthalmol 12:327–331, 1973.

19. Feke GT, Tagawa H, Deupree DM, Goger DG, Delori FC, Weiter JJ: Laser Doppler measurement of regional blood flow in the normal

human retina. Invest Ophthalmol Vis Sci [ARVO Suppl]:224, 1985.

20. Weiter JJ: Studies on the retinal circulation and oxygen transport to the retina. PhD Thesis, University of Chicago, 1979.

21. Warburg O, Posener K, Negelin E: Uber den Stoffwechsel der Carcinomcell, Biochem Z 152:309, 1924.

22. Zuckerman R, Weiter JJ: Oxygen transport in the bullfrog retina. Exp Eye Res 30:117–127, 1980.

23. Weiter JJ, Zuckerman R: The influence of the photoreceptor-RPE complex on the inner retina. An explanation for the beneficial effects of photocoagulation. Ophthalmology 87:1133–1139, 1980.

24. Tillis TN, Schmidt GJ, Weiter JJ: In vivo light and dark oxygen measurements under normoxic conditions in the avascular rabbit retina. Invest Ophthalmol Vis Sci [ARVO Suppl] 27:318, 1986.

25. Feke GT, Zuckerman R, Green GJ, Weiter JJ: Response of human retinal blood flow to light and dark. Invest Ophthalmol Vis Sci 24:136–141, 1983.

26. Young RW: Biological Renewal. Applications to the eye. Trans Ophthalmol Soc UK 102:42–61, 1982.

27. Wing GL, Blanchard GC, Weiter JJ: The topography and age relationship of lipofuscin concentration in the retinal pigment epithelium. Invest Ophthalmol Vis Sci 17:601–607, 1978.

28. Weiter JJ, Delori FC, Wing GL, Fitch KA: Retinal pigment epithelium lipofuscin and melanin and choroidal melanin in human eyes. Invest Ophthalmol Vis Sci 27:145–152, 1986.

29. Lerman S: Radiant energy and the eye. Macmillan Publishing Co, New York, p 153, 1980.

30. Feeney-Burns L, Berman ER, Rothman H: Lipofuscin of human retinal pigment epithelium. Am J Ophthalmol 90:783, 1980.

31. Feeney-Burns L, Hilderbrand ES, Eldridge S: Aging human RPE: morphometric analysis of macular, equatorial and peripheral cells. Invest Ophthalmol Vis Sci 25:195, 1984.

32. Katz ML, Robinson WG Jr: Age-related changes in the retinal pigment epithelium of pigmented rats. Exp Eye Res 38:137, 1984.

33. Weiter JJ, Fine BS: A histologic study of regional choroidal dystrophy. Am J Ophthalmol 83:741–750, 1977.

34. Feeney-Burns L: The pigments of the retinal pigment epithelium, in Current Topics in Eye Research, vol 2, Zadunaisky JA, Davson H (eds). Academic Press, New York, pp 119–178, 1980.

35. Weiter JJ, Delori FC, Wing GL, Fitch KA: Relationship of senile macular degeneration to ocular pigmentation. Am J Ophthalmol 99:185–187, 1985.

36. Garcia RI, Szabo G, Fitzpatrick TB: Molecular and cell biology of melanin, in The Retinal Pigment Epithelium. Zinn KM, Marmor MF (eds). Harvard University Press, Cambridge, MA, pp 124–147, 1979.

37. Rapp LM, Williams TP: The role of ocular pigmentation in protecting against retinal light damage. Vis Res 20:1127–1131, 1980.

38. Krinsky NI: The protective function of carotenoid pigments, in Photo-

physiology, Giese AC (ed). vol 3. Academic Press, New York, pp 123–195, 1982.

39. Menon IA, Hakerman HF: Mechanisms of action of melanins. Br J Dermatol 97:109–112, 1977.

40. Feeney-Burns L, Berman ER: Oxygen toxicity: membrane damage by free radicals. Invest Ophthalmol 15:789, 1976.

41. Barr FE, Saloma JS, Buchele MJ: Melanin: the organizing molecule. Med Hypothesis 11:1, 1983.

42. Hunold W, Malessa P: Spectrophotometric determination of the melanin pigmentation of the human ocular fundus in vivo. Ophthalmic Res 6:355, 1974.

43. Wald G: Human vision and the spectrum. Science 101:653, 1945.

44. Malinow MR, Feeney-Burns L, Peterson LH, et al: Diet-related macular anomalies in monkeys. Invest Ophthalmol Vis Sci 19:857–873, 1980.

45. Snodderly DM, Brown PK, Delori FC, Auran JD: The macular pigment. I. Absorbance spectra, localization, and discrimination from other yellow pigments in primate retinas. Invest Ophthalmol Vis Sci 25:660–673, 1984.

46. Reading VM, Weale RA: Macular pigment and chromatic aberration. J Opt Soc Am 64:231, 1974.

47. Lawwill T, Crockett RS, Currier G: The nature of chronic light damage to the retina, in Williams TP, Baker BN (eds). The effects of constant light on visual processes. Plenum Press, New York, pp 161–177, 1980.

48. Bone RA, Sparrock JMB: Comparison of macular pigment densities in human eyes. Vision Res 11:1057, 1971.

49. Hayes KC: Retinal degeneration in monkeys induced by deficiencies of vitamin E or A. Invest Ophthalmol 13:499–510, 1974.

50. Farnsworth CC, Dratz EA: Oxidative damage of retinal rod outer segment membranes and the role of vitamin E. Biochem Biophys Acta 443:556–570, 1976.

51. Katz ML, Parker KR, Handelman GJ, Barnel TL, Dratz EA: Effects of antioxidant nutrient deficiency on the retina and retinal pigment epithelium of albino rats: a light and electron microscopic study. Exp Eye Res 34:329–369, 1982.

52. Weiter JJ, Dratz EA, Fitch K, Handelman G: Role of selenium nutrition in senile macular degeneration. Invest Ophthalmol Vis Sci 26[ARVO Suppl]:58, 1985. (abstract)

53. Tso MOM, La Piana FG: The human fovea after sungazing, Trans Am Acad Ophthalmol Otolaryngol 79:788, 1975.

54. Hamm WT, Mueller HA, Ruffolo JJ, Guerry D: Solar retinopathy as a function of wavelength: Its significance for protective eyewear, in the effects of constant light on visual processes. Williams TP, Baker BN (eds). Plenum Press, New York, pp 319–346, 1980.

55. Cain CP, Welch AJ: Measured and predicted laser-induced temperatue rises in the rabbit fundus. Invest Ophthalmol 13:60–70, 1974.

56. Hogan MJ, Alvarado JA, Weddell JE: Histology of the human eye.

An Atlas and Textbook. WB Saunders Co, Philadelphia, p 422, 1971.

57. Kahn HA, Moorhead HB: Statistics on blindness in the model reporting area. Department of Health, Education and Welfare, Publication No. (NIH) 73–427 (1969–1970).

58. Burns RP, Feeney-Burns L: Clinico-morphologic correlations of drusen of Bruch's membrane. Trans Am Ophthalmol Soc 78:206–225, 1980.

59. Weiter JJ, Jalkh A, Trempe C, Pruett R: Management of retinal pigment epithelial detachment in senile macular degeneration, in Modern Concepts in Vitreo-Retinal Diseases, Adolphe Neetens MD (ed). University of Antwerp UIA Press, pp 119–124, 1985.

60. Tucker MA, Shields JA, et al: Sunlight exposure as risk factor for intraocular malignant melanoma. N Engl J Med 313:789–792, 1985.

61. Lew RA, Sober AJ: Sun exposure habits in patients with cutaneous melanoma: a case study. J Dermatol Surg Oncol 9:98–106, 1983.

62. Weiter JJ, Delori FC: An explanation for the "Bull's Eye" macular lesion. Invest Ophthalmol Vis Sci [ARVO Suppl] 27:336, 1986.

63. Terry TL: Extreme prematurity and fibroplastic overgrowth of persistent vascular sheath behind each crystalline lens. I. Preliminary report. Am J Ophthalmol 25:203, 1942.

64. Owens WC, Owens EU: Retrolental fibroplasia in premature infants. II. Studies on the prophylaxis of the disease: the use of alpha tocopheryl acetate. Am J Ophthalmol 32:1631–1637, 1949.

65. Kinsey VE, Zacharias L: Retrolental fibroplasia: incidence in different localities in recent years and a correlation of the incidence with treatment given the infants. JAMA 139:572–578, 149.

66. Campbell K: Intensive oxygen therapy as a possible cause of retrolental fibroplasia: a clinical approach. Med J Aust 2:48–50, 1951.

67. Patz A, Hocck LE, De LaCruz E: Studies on the effect of high oxygen administration in retrolental fibroplasia. I. Nursery observations. Am J Ophthalmol 35:1248–1253, 1952.

68. Ashton N, Ward B, Sergell G: role of oxygen in the genesis of retrolental fibroplasia: a preliminary report. Br J Ophthalmol 37:513–520, 1953.

69. Kinsey VE, Hemphill FM: Etiology of retrolental fibroplasia and preliminary report of cooperative study of retrolental fibroplasia. Trans Am Acad Ophthalmol Otolaryngology 59:15–24, 1955.

70. Weiter JJ, Zuckerman R: The influence of the photoreceptor—RPE complex on the inner retina: an explanation for the beneficial effects of photocoagulation. Ophthalmology 87:1133–1139, 1980.

71. Weiter JJ, Zuckerman R, Schepens CL: A model for the pathogenesis of retrolental fibroplasia based on the metabolic control of blood vessel development. Ophthalmol Surg 13:1013–1017, 1982.

72. Ashton N: Oxygen and the growth and development of retinal vessels, in Kimura SJ, Coygill WN (eds). Vascular Complications of Diabetes Mellitus. CV Mosby, St Louis, pp 3–32, 1967.

73. Crapo JD, McCord JM: Oxygen-induced changes in pulmonary superoxide dismutase assayed by antibody titrations. Am J Physiol 231:1196–1203. 1976.

74. Hittner HU, Godio LB, Rudolph AJ, et al: Retrolental fibroplasia: effect of vitamin E in a double-blind clinical study of preterm infants. N Engl J Med 305:1365–1371, 1981.

75. Weiter JJ: Retrolental fibroplasia: an unresolved problem. N Engl J Med 305:1404–1406, 1981.

76. Glass P, Avery G, et al: Effect of bright light in the hospital nursery on the incidence of retinopathy of prematurity. N Engl J Med 313:401–404, 1985.

77. Reeser FH, Aaberg TM: In Physiology of the Human Eye and Visual System, Records RE (ed). Harper and Row, Hagerstown, pp 261–295, 1979.

78. Balays EA: In Chemistry and Molecular Biology of the Intercellular Matrix, Balays EA (ed). vol 1. Academic Press, New York, p 293, 1970.

79. Andley UP, Chakrabarti B: Role of singlet oxygen in the degradation of hyaluronic acid. Biochem Biophysic Res Comm 115:894–901, 1983.

7

Light-Induced Changes in the Skin of the Lid

Jeffrey D. Bernhard

Sometime too hot the eye of heaven shines
And often is his gold complexion dimm'd . . .

Shakespeare
Sonnet 18

Introduction

Like the rest of the integument, the eyelids and periorbital skin are subject not only to normal acute and chronic effects of ultraviolet radiation, but to abnormal reactions as well. Such changes range from clinically inapparent molecular damage to sunburn during the acute phase, and from the ravages of sun-induced aging such as solar elastosis to basal cell carcinoma over the long term. Of the many dermatologic diseases that are exacerbated by sunlight, such as lupus erythematosus and chemical photocontact dermatitis, most can and do involve the lids and periorbital skin on occasion. The normal open position of the eye during most outdoor activities (except for unprotected sunbathing), the eyebrows, and the anatomically depressed and therefore partially shaded position of the orbit, together offer some protection to the periorbital skin, especially the upper eyelid. In fact, eyelid and periorbital sparing are often taken as clinical clues that light exposure may be an important contributory factor to or the direct cause of a cutaneous eruption that does involve other parts of the face.

The study of normal and abnormal cutaneous reactions to light comprises the core of clinical photomedicine. The first eight references are monographic sources of information on this subject, highlights of which are summarized in this chapter, but which the reader is urged to consult for more detailed information and primary citations.[1-8]

Historically, the study of light-induced changes of the skin is closely related to the study of cutaneous aging, but the two are not, in fact, the same. Confusion between intrinsic aging and the process of chronic sun damage has led the effects of the latter to be erroneously labeled as aging, premature aging, or accelerated aging.[3] Although the ultimate cumulative effects of chronic sun exposure may indeed make a person look older than his chronological age, it is important to separate effects due to time from those due to sunlight. In Sonnet 60, Shakespeare gave an accurate clinical description of the aging face: "Time doth transfix the flourish set on youth, and delves the parallels in beauty's brow . . ." However, he misattributed the blame and perhaps ought to have said "solar radiation" for "Time." New findings on chronic sun damage, which would have improved Shakespeare's accuracy but perhaps not his poetry, are discussed in this chapter.

Acute Responses of Normal Skin to Ultraviolet Radiation: The Human Sunburn Reaction [7–9]

Although solar ultraviolet (UV) radiation of wavelengths between 290 to 320 nm (UV-B) reaches the earth in relatively small quantities, it is very efficient in causing sunburn of human skin. Considerably more UV-A radiation (320–400 nm) reaches the earth's surface, but it is orders of magnitude less effective in producing the delayed erythemal response known as sunburn.[8] Ultraviolet-B and UV-A are both erythemogenic and melanogenic, but owing to their relative photobiologic efficiency, it is predominantly the UV-B that leads to the painful aftereffects of an inadvertent snooze in the sun, although UV-A may contribute as much as 15% of erythemally effective radiation at noon.[8]

Since sun-exposure habits are now surrounded by no less mythology than the sun itself was in ancient times, it is probably worth commenting on several popular misconceptions. Clouds do not prevent sunburn; up to 80% of UV radiation "penetrates" them. Since UV radiation is reflected off sand, water, and porch decks, beach umbrellas do not, in fact, provide full protection. Ultraviolet radiation penetrates water and can lead to a severe burn.

The sequence of photochemical, molecular, and cellular events set in motion by the interaction of UV photons with human skin is not known in precise detail, although recent work has led to considerable insight.[7] In addition to the well-known formation of DNA photoproducts such as pyrimidine dimers, photochemical alterations of RNA, of structural and enzymatic proteins, and of membranes occur as well. These changes, in turn, may lead to impairment of cell function, repair processes, and sequential inflammatory events.[7]

At the histologic level, the human sunburn reaction is marked by the appearance of dyskeratotic and vacuolated keratinocytes ("sunburn cells") in the epidermis.[9] In addition, specialized antigen-presenting dendritic (Langerhans) cells disappear,[9] whereas antigen-presenting cells that lead to immunologic suppression, and which are resistant to UV radiation (the Granstein cells), remain.[9a] Major changes in the dermis include endothelial cell enlargement in both superficial and deep venular plexuses. The earliest events include mast cell degranulation and associated perivenular edema, which are apparent by one hour and maximal three to four hours after irradiation, when erythema begins to appear. By 24 hours the edema subsides, the mast cells return to normal number and granule content, and erythema peaks. At the biochemical level, prostaglandin E_2 (PGE_2) begins to rise even before the onset of erythema, while histamine levels rise to approximately fourfold above baseline values just after onset of erythema and return to normal within 24 hours.[9] Other inflammatory mediators, such as kinins, and cell types, such as the polymorphonuclear leukocyte, have also been implicated in the induction of UV radiation-induced inflammation (reviewed in references 4 and 7).

Clinically, erythema is the most notable cutaneous response to UV radiation, and it may be caused by a single, sufficient dose of appropriate UV wavelength. The smallest dose of energy required to produce an area of erythema with distinct margins is defined as the minimal erythema dose (MED). It is expressed as energy delivered per unit area (e.g., millijoules or joules per square centimeter).

Depending on the wavelength, exposure to UV radiation doses progressively larger than the MED will lead to the classic signs of inflammation: tumor (swelling), calor (warmth), and dolor (pain), which may last for hours or days. These are followed by epidermal hyperproliferation, hyperpigmentation, and, frequently, desquamation.[7] Even at suberythemogenic exposure doses, however, DNA damage, epidermal cell death, and abnormal differentiation occur.[10-12] Finally, there is clear evidence that a single suberythemogenic exposure to UV radiation influences the response to subsequent exposures[13] and that repeated suberythemogenic exposures have cumulative effects.[14-15] These findings, together with the evidence that photocarcinogenesis, vessel injury, and other clinically inapparent effects may have action spectra that differ from the action spectrum for delayed erythema, have led to John Parrish's important warning that "delayed erythema cannot be used as the only indicator of phototoxicity (cell injury by photons)."[7] Lastly, it is important to note that the time-course, dose-response curve, histopathologic changes, and probably biochemical changes as well, are different for UV-A, UV-B, and UV-C (200–290 nm).[7,16]

Benign Changes of the Skin Induced by Chronic UV Radiation Exposure: Dermatoheliosis

Chronic sun exposure leads to a variety of cumulative and irreversible effects on the skin's vasculature, keratinocytes, melanocytes, and connective tissue. The sum total of these changes leads to a clinical syndrome that Fitzpatrick has labeled "dermatoheliosis."[1] It includes telangiectasia, atypical keratinocytic hyperplasia (actinic keratosis), freckles, solar lentigines, wrinkling, roughening, and yellowing of the skin.[1] Sunlight, not innate aging, is the main cause of the worst features of senile skin. A variety of genetic factors, including melanogenic capability, probably determine susceptibility to these changes. Although no race is spared, it is fair-skinned caucasians with a history of significant sun exposure who exhibit dermatoheliosis most severely. The face, neck, and extensor surfaces of the upper extremities are affected most often.[17–19] The eyelids and periorbital skin are affected to the extent that a susceptible individual's sun-exposure habits, during work and recreation, subject these areas to solar radiation. To some extent a shade or shielding effect is exerted by the depressed position of the orbit and protuberance of the brow. But these can easily be subverted by reflection from environmental surfaces and by intentional sunbathing, particularly when protective goggles are not used or when artificial reflectors are.

The most important features of dermatoheliosis (actinically damaged skin) are summarized in Table 7.1. Of the clinical abnormalities enumerated, wrinkling and the *maladie de Favre et Racouchot* (nodular cutaneous elastoidosis with cysts and comedones) are particularly noteworthy for their localization to temporal and periorbital areas, particularly the infraorbital ridge (Fig. 7.1). This condition is a slowly progressive but benign disorder that occurs mainly in older men who have had extensive exposure to sun and weather.[20] Its most distinctive feature is the presence of large comedones that, in combination with the associated features of furrows, nodularity, atrophic changes, and a yellowish to brown hue, are so characteristic that the diagnosis can be made clinically. When in doubt, histopathologic analysis of a biopsy specimen will reveal thinning of the epidermis, flattening of the rete ridges, cysts, and basophilic degeneration of collagen (solar elastosis) in the dermis separated by a narrow band of normal collagen from the overlying epidermis.[20] Therapies that may be helpful in treatment of the various components of dermatoheliosis are listed in Table 7.2.

Table 7.1. Features of actinically damaged skin.[*]

Clinical abnormalities	Histologic abnormality	Presumed pathophysiology
Epidermis		
Dryness (roughness)	Minimal stratum corneum irregularity	Altered keratinocyte maturation
Actinic keratoses	Nuclear atypia; loss of orderly, progressive keratinocyte maturation; irregular epidermal hyper- and/or hypoplasia; occasional dermal inflammation	Premalignant disorder
Irregular pigmentation		
Freckling	Reduced number of hypertrophied, strongly dopa-positive melanocytes	Reactive hyperplasia and later loss of functional melanocytes
Lentigines	Elongation of epidermal rete ridges; increase in number and melanization of melanocytes	
Guttate hypomelanosis	Absence of melanocytes	
Dermis		
Wrinkling		
Fine surface lines Deep furrows	None detected	Alterations in dermal matrix and fibrous proteins
Stellate pseudoscars	Absence of epidermal pigmentation, altered dermal collagen	Loss of functional melanocytes, reactive collagen deposition by fibroblasts
Elastosis (fine nodularity and/or coarseness)	Nodular aggregations of fibrous to amorphorous material in the papillary dermis	Overproduction of abnormal elastin fibers
Inelasticity	Elastotic dermis	Altered elastin fibers
Telangiectasia	Ectatic vessels often with atrophic walls	Loss of connective tissue support
Venous lakes	Ectatic vessels often with atrophic walls	Loss of connective tissue support
Purpura (easy bruising)	Extravasated erythrocytes	Loss of connective tissue support for dermal vessel walls
Appendages		
Comedones (maladie de Favre et Racouchot)	Ectatic superficial portion of the pilosebaceous follicle	Loss of connective tissue support
Sebaceous hyperplasia	Concentric hyperplasia of sebaceous glands	Increased mitotic and functional responsiveness of glandular tissue

[*] Reprinted with permission from Gilchrest BA: Skin and Aging Processes. Copyright CRC Press, Inc., Boca Raton, Fl.

Biologically speaking, it is important to distinguish changes caused by intrinsic aging from changes caused by chronic sunlight exposure, even though sunlight provides the major contribution

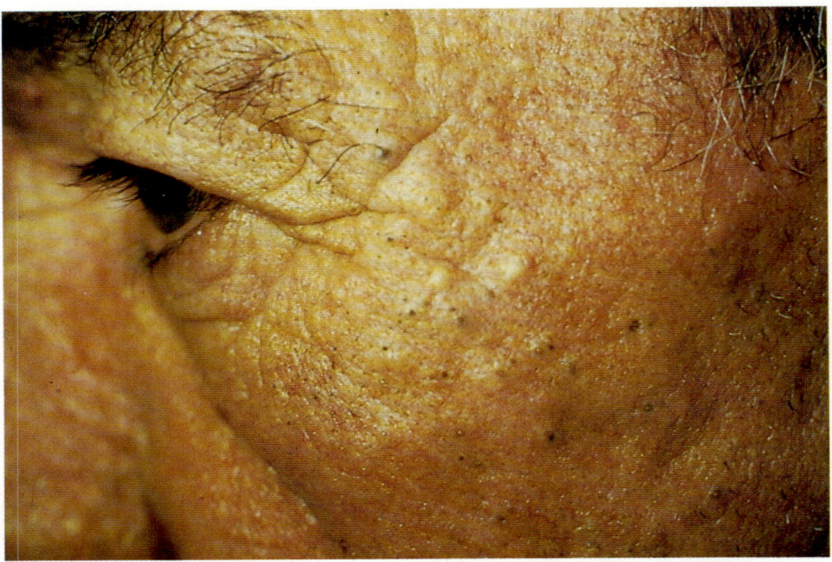

Figure 7.1. Maladie de Favre et Racouchot (nodular cutaneous elastoidosis with cysts and comedones): a consequence of severe, long-term actinic damage. (Courtesy of Dr. G. Bishop.)

Table 7.2. Treatment of dermatoheliosis.[*]

Component lesions	Effective therapies
Dryness (roughness)	Emollients (transient effect)
Actinic keratoses	Topical fluorouracil; cryotherapy/dermabrasion/chemical peel trichloroacetic acid (TCA)
Freckling	Hydroquinone (2%–5%)[†]
Lentigines	Cryotherapy; hydroquinone[†]
Guttate hypomelanosis	NR[‡]
Wrinkles	
Fine surface lines	Chemical peel (TCA)
Deep furrows	Rhydidectomy (face lift); collagen injections
Stellate pseudoscars	NR
Elastosis	NR
Inelasticity (redundant skin)	Rhydidectomy
Telangiectasia	Electrocautery; argon laser
Venous lakes	NR
Purpura	NR
Comedones	Retin-A® (tretinoin; all-trans retinoic acid); manual expression
Sebaceous hyperplasia	Excision

[*] Reprinted with permission from Gilchrest BA: Skin and Aging Processes. Copyright CRC Press, Inc., Boca Raton, Fl.
[†] Minimally effective for most patients.
[‡] NR = None reported.

to the *clinical* appearance of aging. One need only compare the sun-exposed skin of the face of a 70-year-old man to the unexposed skin of his buttock to be convinced that sunlight is a major culprit in this regard. Although certain changes such as flattening of the epidermis[21] and degradative alterations of the elastic fiber network[22] occur with increasing age in both sun-exposed and nonexposed skin, other changes are clearly more pronounced in the former. Examples include production of excessive basement membrane-like material by veil cells of the cutaneous microvasculature[23] and scar-like collagen deposition in the papillary dermis of actinically damaged skin[24] (see also reference 25).

A variety of pigmentary changes and specific pigmented lesions occur with great frequency in sun-damaged white skin. Solar damage leads to reactive hyperplasia, depletion of injured melanocytes, and impaired transfer of melanin pigment via melanosomes to keratinocytes.[3] Freckles (ephelides) contain a reduced number of melanocytes, but those present are larger, more dendritic, and more heavily pigmented than normal. So-called "senile" or "solar" lentigines (also referred to by laypersons as "age spots" and "liver spots") are characterized by elongation of the epidermal rete ridges and increased pigment production by an increased number of melanocytes. These are macular (flat) tan to brown benign lesions without significant surface, border, or color irregularity.[26] In contrast,

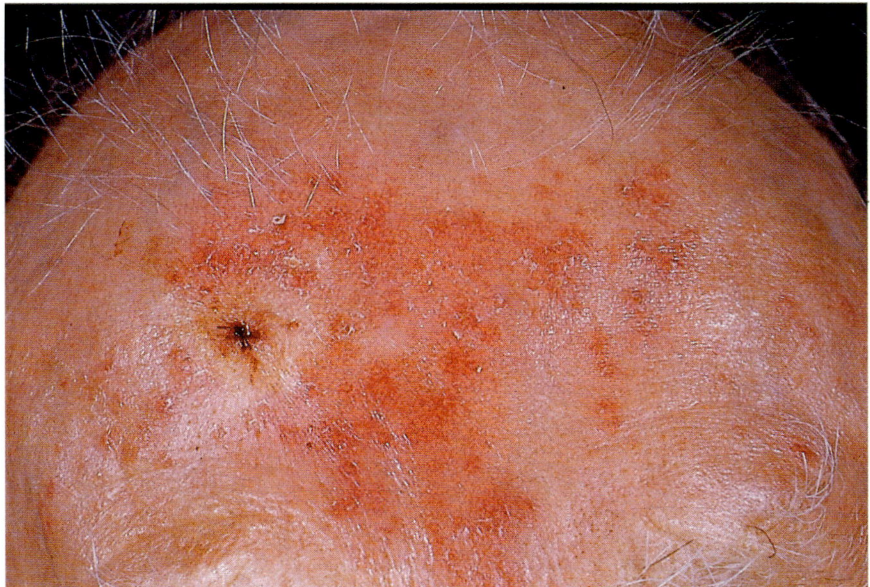

Figure 7.2. Actinic keratoses on the forehead. A suture is in place at site of biopsy that was performed to exclude squamous cell carcinoma.

the lentigo maligna is a precursor to malignant melanoma. It is characterized histologically by atypical melanocytic hyperplasia, and clinically as a hyperpigmented flat lesion—almost exclusively on exposed skin—that develops variegation of color and border.[27,28]

Actinic keratoses are extremely common in actinically damaged skin (Fig. 7.2). The lesions are small, erythematous, scaling plaques that may be almost flat but can become very hyperkeratonic. The scale often has a gritty, sandpaper quality. Histologically, there is nuclear atypia, disorderly keratinocyte maturation, irregular epidermal hyperplasia or hypoplasia, and occasionally dermal inflammation.

Malignant Changes of the Skin Induced by Chronic UV Radiation Exposure: Photocarcinogenesis

The importance of sunlight as an etiologic factor in the development of skin cancer in man has been recognized since around the turn of the 20th century.[29] In the absence of exogenous photosensitizing agents, the action spectrum for these effects appears to lie primarily within the UV-B (290–320 nm) range.[29,30] The most impressive associations between sunlight and human skin cancer relate to basal cell and squamous cell carcinomas (Figs. 7.3–7.7). Clinically, this is most apparent in the striking localization

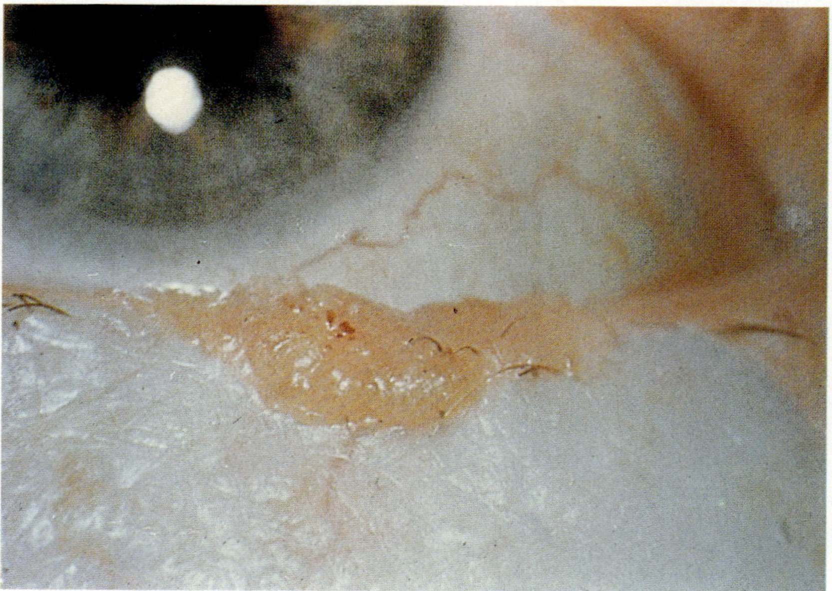

Figure 7.3. Basal cell carcinoma on lower eyelid. (Courtesy of Dr. G. Bishop.)

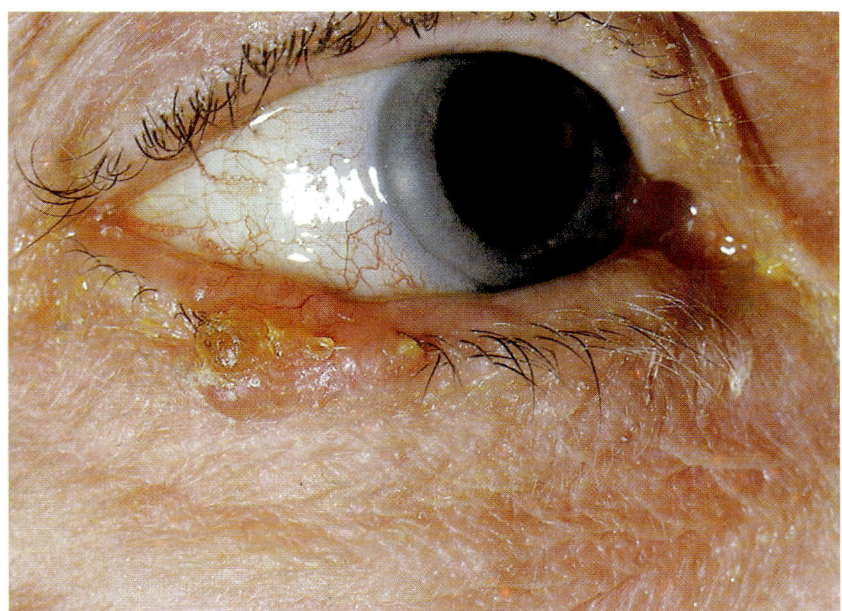

Figure 7.4. Basal cell carcinoma on lower eyelid. (Courtesy of Drs. J. Gittinger and H.A. Kachadoorian.)

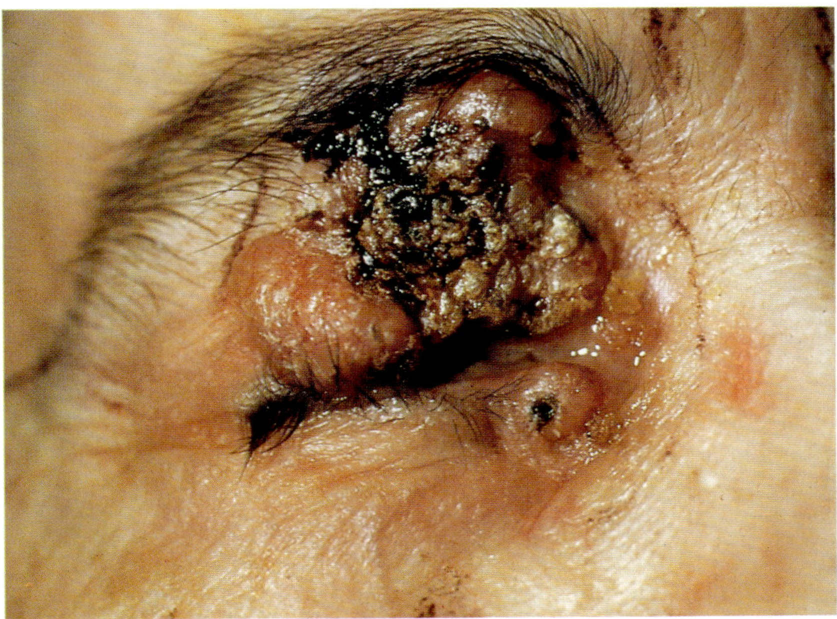

Figure 7.5. Recurrent basal cell carcinoma. (Courtesy of Dr. G. Bishop.)

of these tumors to sun-exposed skin. The association between sunlight and malignant melanoma has been more controversial, but the weight of accumulated indirect evidence can leave hardly any doubt of its importance.[31,32]

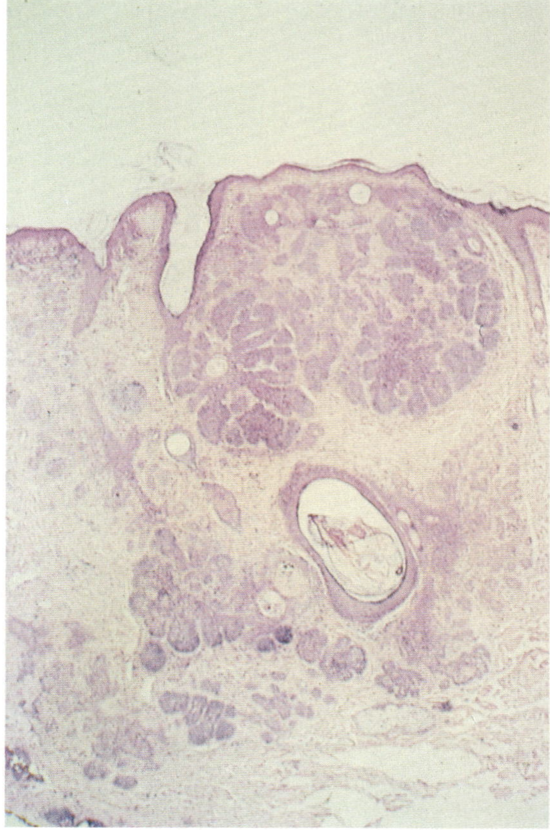

Figure 7.6. Histopathology of basal cell carcinoma. Note islands of hyperchromatic basaloid tumor cells. (Courtesy of Dr. R. Malhotra.)

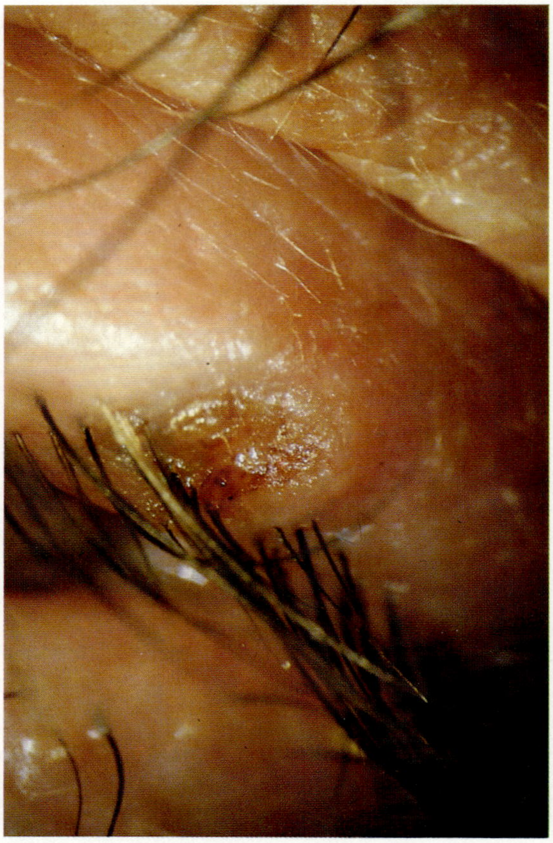

Figure 7.7. Squamous cell carcinoma of upper eyelid. (Courtesy of Drs. J. Gittinger and H.A. Kachadoorian.)

The mechanism of photocarcinogenesis has been reviewed in detail elsewhere.[29,30] It probably involves the interplay of a number of variables, including genetically determined susceptibility, direct photochemical effects of UV radiation on DNA and other macromolecules, and complex photoimmunologic events.[33]

Basal cell carcinoma is the most common malignant tumor of the skin and accounts for 70% to 75% of all skin tumors. White-skin people with light hair, light eyes, and fair complexions are at the greatest risk. The nose, eyelids, cheeks, and trunk are the most common sites. Metastases are extremely rare, but can occur. Given the comparative degrees of exposure to sunlight, it is not surprising that among eyelid basal cell carcinomas, the lower eyelid is the most common site, followed in most series by the inner canthal area, upper lid, and outer canthal region.[34]

Basal and squamous cell carcinomas near the eye are obviously dangerous and do enter the orbit on occasion, particularly when recurrences occur.[34] Near the canthi, especially the inner canthi, basal cell carcinomas can be extremely difficult to detect. Any individual with a history or evidence of significant sun exposure should be examined carefully for basal cell cancer near the eyes under adquate lighting, preferably in the supine position.

Actinic keratoses deserve careful attention anywhere on the skin, but especially on the face and near the eye. Not only do they have the potential to develop into squamous cell carcinomas, but they alert the clinician that sun exposure in the past has been significant enough to warrant close examination, treatment when indicated, and careful long-term follow-up for possible occurrence of basal and squamous cell carcinomas.

Treatment options for skin cancer include surgical excision, curettage with electrodessication, radiation, cryosurgery, and microscopically controlled chemosurgery. For the upper eyelids, surgical excision is often ideal because the skin's laxity permits primary closure.[35] Surgical excision is also recommended for the lower eyelids, but it is important to stress that adequate surgical margins be obtained: recurrences can be very difficult to eradicate.[35] Since treatment failure in the canthi can lead to orbital involvement, careful selection of the treatment modality is of the utmost importance. Microscopically controlled chemosurgery (Moh's technique) may be the treatment of choice, especially for poorly defined tumors and recurrences.[35] The cooperation of a dermatologic chemosurgeon, plastic surgeon, and ophthalmologist may be required. It is not unusual to find that the microscopic spread of tumor far exceeds what is anticipated in clinical examination of the tumor.

The differential diagnosis of tumors of the eyelids includes sebaceous carcinoma, other malignant tumors, benign tumors, cysts, and infections, and benign lesions that may be clinically and even histologically confused with cancer. Of the last, keratoacanthoma is important because of its similarity to squamous cell carcinoma; trichoepithelioma because of its similarity to basal cell carcinoma.[34]

Photodermatoses and Other Skin Diseases Provoked or Exacerbated by UV Radiation [6]

The skin of the lids and immediate periorbital area is susceptible to a large number of cutaneous disorders that may be provoked or exacerbated by light. These are summarized in Tables 7.3 and 7.4. Of the genetic and metabolic disorders, the porphyrias are

Table 7.3. Classification of abnormal cutaneous reactions to light in man.[*]

  I. Genetic and metabolic disorders
     A. Light alone
            Ephelides (freckles)[†]
            Xeroderma pigmentosum[†]
            Bloom's syndrome[†]
            Cockayne's syndrome[†]
            Rothmund's syndrome[†]
            Melanin deficiency syndromes[‡]
               Albinism
               Phenylketonuria
               Vitiligo
               Hypomelanotic individuals (skin type I)
     B. Light plus endogenous metabolite
            Certain porphyrias[†]
            Disorders of tryptophan metabolism[†]
               Hartnup syndrome
               Hydroxykynurenuria
               Carcinoid syndrome
               Pellagra
     C. Light plus exogenous chemical[†]
            Certain porphyrias
               Hexachlorobenzene-induced porphyria turcia
               Alcohol- or estrogen-induced porphyria
               cutanea tarda
 II. Idiopathic photodermatoses[†]
     A. Acute intermittent photodermatoses (occurring or re-occurring within
        minutes to hours of single exposures)
            Polymorphic light eruption and variants
            Solar urticaria
     B. Chronic persistent photodermatoses (chronic actinic dermatitis; acute
        exacerbations may occur)
            Photosensitive eczema
            Actinic reticuloid
III. Chemical photosensitivity
     A. Phototoxicity[‡]
     B. Photoallergy[†]
     C. Persistent light reactivity[†,‡]
     D. Phytophotodermatitis[†]
     E. Certain genetic and metabolic disorders[†,‡]
 IV. Degenerative and neoplastic disorders[†]
     A. Dermatoheliosis
     B. Actinic keratoses
     C. Stucco keratoses
     D. Granuloma solare
     E. Bowen's disease
     F. Squamous cell carinoma
     G. Basal cell carcinoma
     H. Melanoma
     I. Idiopathic guttate hypomelanosis

[*] Reproduced with permission from Bernhard JD, Parrish JA, Pathak MA,

Table 7.4. Disorders that can be precipitated, provoked, or exacerbated by light.[*]

Acne vulgaris
Atopic eczema
Bullous pemphigoid
Darier-White disease
Erythema multiforme
Herpes simplex labialis (recurrences)
Lichen planus
Lupus erythematosus
Pemphigus erythematosus
Pemphigus foliaceous
Pemphigus vulgaris
Hailey-Hailey disease
Physical occlusion of skin (increased susceptibility to sunburn)
Pityriasis alba
Pityriasis rubra pilaris
Psoriasis
Reticular erythematous mucinosis syndrome
Rosacea
Seborrheic dermatitis
Transient acantholytic dermatosis (Grover's disease)
Viral infections of the skin
Vitiligo

[*] Reproduced with permission from Bernhard JD, Parrish JA, Pathak MA, Kochevar IE: Abnormal responses to ultraviolet radiation, in Dermatology in General Medicine, ed 3, Fitzpatrick TB, Eisen AZ, Wolff K, Freedberg IM, Austen KF (eds). McGraw-Hill, New York, p. 1482, 1987.

frequently associated with skin changes near the eye and are discussed below. Of the idiopathic photodermatoses, polymorphic light eruption frequently spares the face, perhaps because of "desensitization" due to chronic ambient UV radiation. Phototoxicity and photoallergy frequently involve the skin of the face and lids and may lead to massive periorbital edema. Phototoxic reactions can occur in any individual exposed to the combination of a sufficient dose of appropriate wavelengths of radiation and a sufficient dose of a photosensitizing chemical or drug. Examples of such chemicals that may be administered orally include demeclocycline

Kochevar IE: Abnormal responses to ultraviolet radiation, in Dermatology in General Medicine, ed 3, Fitzpatrick TB, Eisen AZ, Wolff K, Freedberg IM, Austen KF (eds). McGraw-Hill, New York, p. 1482, 1987.

[†] Usually a qualitatively abnormal response; e.g., papules, plaques, wheals, eczematous lesions, vesicles.

[‡] Usually a quantitatively abnormal response, as in lowered erythema threshold. Morphology is that of sunburn; confluent erythema that may progress to include edema and vesiculation.

and 8-methoxypsoralen; topical photosensitizers that lead to photo-toxicity include certain tar derivatives and halogenated salicylani-lides. Photoallergy involves a reaction to a chemical plus radiation in which the individual's immune response also plays a role. Whereas phototoxic reactions most often appear as exaggerated sunburn, photoallergic reactions may consist of an eczematous eruption or of discrete papules and plaques. Because they are so thin and because eyelids are touched so often and are so frequently subjected to application of exogenous chemicals in cosmetics, con-tact dermatitis and photoallergic and phototoxic contact dermatitis are not uncommon.

Cosmetics may contain a variety of photosensitizing chemicals such as essential oils, coal-tar derivatives, and fluorescein deriva-tives. The classic example is 5-methoxypsoralen contained in berga-mot oil; its presence in perfume has led to a syndrome so distinctive as to be dignified by a name of its own—berlocque dermatitis. Wherever the offending perfume, cologne, or aftershave has been dabbed on and exposed to a sufficient dose of long UV rays (UV-A: 320–400 nm), a phototoxic response characterized by erythema, edema, and occasionally vesiculation may occur. Under some cir-cumstances the acute reaction may be mild enough to escape no-tice, only to be followed later by dense hyperpigmentation.

Similar photosensitivity reactions may occur following exposure to plant substances that may contain furocoumarins, especially from plants of the Umbelliferae and Rutaceae orders. After expo-sure to sunlight, *phytophotodermatitis* occurs as a phototoxic reac-tion that may range from mild erythema to severe blistering. It usually appears within several hours to a day after exposure.

Light-induced changes of the eyelids and periorbital skin may be indicative of serious systemic illnesses such as lupus erythema-tosus (LE) and the porphyrias. In LE, acute photosensitivity reac-tions may occur, as may chronic discoid lesions with atrophic and pigmentary changes, scarring, follicular plugging, and hair loss. Figure 7.8 shows the scarred eyelids of a man with discoid cutane-ous LE. Figure 7.9 shows characteristic lesions on the face.

Ocular and periorbital manifestations may occur in porphyria cutanea tarda, congenital erythropoietic porphyria, hereditary co-proporphyria, variegate porphyria, and hepatoerythropoietic por-phyria.[36,37] These may include photophobia, epiphora, blepharo-spasmus, and phlyctenular conjunctivitis.[36,37] Photosensitivity may also lead to blistering of the eyelids and, over the long term, sclerodermoid changes. Trichiasis and madarosis of the eyelashes, and van der Hoeve's scleromalacia perforans may occur as well.[37] Scarring of the lids, keratitis, cataracts, and blindness occur fre-quently in xeroderma pigmentosum.

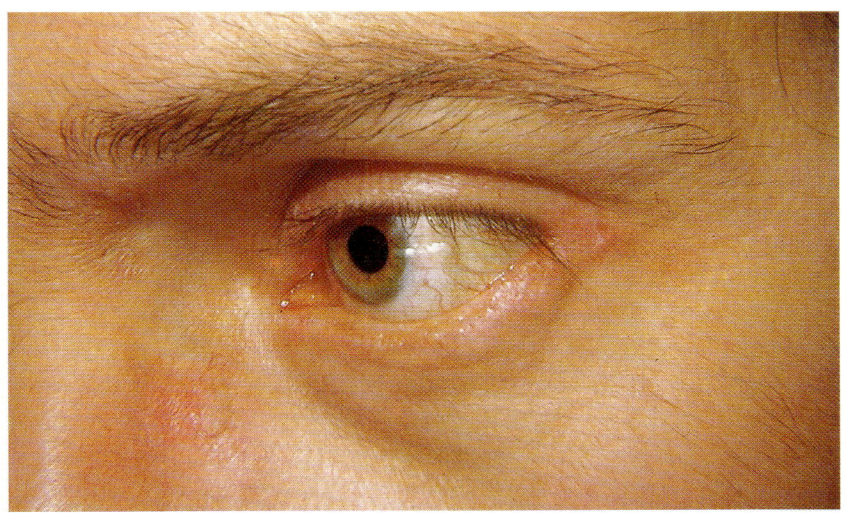

Figure 7.8. Discoid lupus erythematosus affecting lower eyelid. Note erythema, atrophy, pigmentary changes, and scarring loss of lashes.

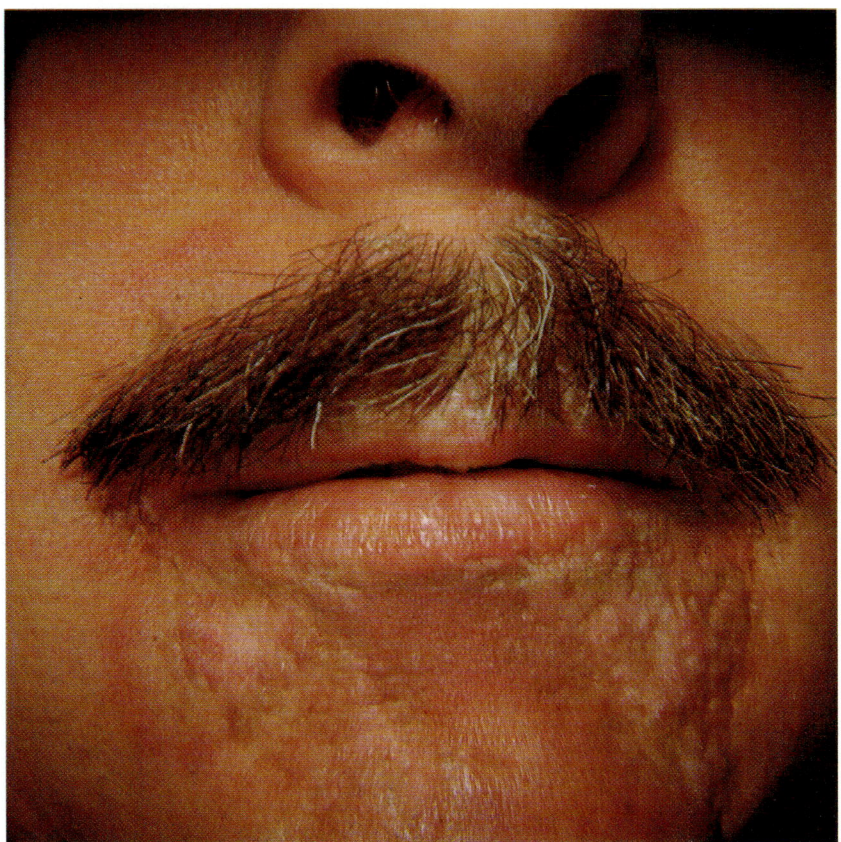

Figure 7.9. Discoid lupus erythematosus. Same patient as in Figure 7.8, with characteristic changes of chronic cutaneous lupus erythematosus on chin and moustache area.

Differential Diagnosis of Skin Changes Near the Eye

A number of skin conditions may be associated with ocular findings, and it is obvious that most diseases of the skin can occur on the face and eyelids as well. Those associated with light exposure have been discussed above. For a thorough treatment of the former, the reader is referred to standard texts and to references 38–40. Although light plays an important role in localizing some skin conditions to the face and eyelids, it may be appropriate to note in closing that chronic low-dose ambient UV radiation exposure has been proposed as the explanation for why at least one skin disease, psoriasis, usually does not involve the face.[41]

References

1. Fitzpatrick TB, Pathak MA, Greiter F, Mosher DB, Parrish JA: Heritable melanin déficiency syndromes, in Update: Dermatology in General Medicine. Fitzpatrick TB, Eisen AZ, Wolff K, Freedberg IM, Austen KF (eds). McGraw-Hill, New York, pp 46–60, 1983.
2. Magnus IA: Dermatological Photobiology. Clinical and Experimental Aspects. Blackwell Scientific Publications, Oxford, 1976.
3. Gilchrest BA: Dermatoheliosis (Sun-Induced Aging), in Skin and Aging Processes. Chapt 8, CRC Press Inc, Boca Raton, pp 97–116, 1984.
4. Hawk JLM, Parrish JA: Responses of normal skin to ultraviolet radiation, in The Science of Photomedicine. Regan JD, Parrish JA (eds). Chapt 8, Plenum Press, New York, pp 219–260, 1982.
5. Harber LC, Bickers DR: Photosensitivity Diseases. Principles of Diagnosis and Treatment. Saunders, Philadelphia, 1981.
6. Bernhard JD, Parrish JA, Pathak MA, Kochevar IE: Abnormal responses to ultraviolet radiation, in Dermatology in General Medicine, ed 3, Chapt. 129, Fitzpatrick TB, Eisen AZ, Wolff K, Freedberg IM, Austen KF (eds). McGraw-Hill, New York, pp 1481–1507, 1987.
7. Parrish JA: Responses of skin to visible and ultraviolet radiation, in Biochemistry and Physiology of the Skin, Goldsmith LA (ed). Chapt 31, Oxford University Press, New York, pp 713–733, 1983.
8. Parrish JA, Anderson RR, Urbach F, Pitts D: UV-A. Biological Effects of Ultraviolet Radiation with Emphasis on Human Responses to Longwave Ultraviolet. Plenum Press, New York, 1978.
9. Gilchrest BA, Soter NA, Stoff JS, Mihm MC: The human sunburn reaction: histologic and biochemical studies. J Am Acad Dermatol 5:411–422, 1981.
9a. Edelson RL, Fink JM: The immunologic function of skin. Scientific American 252:46–53, 1985.
10. Gschnait F, Brenner W, Wolff K: Photoprotective effect of a psoralen-UVA-induced tan. Arch Dermatol Res 263:181–188, 1978.

11. Kaidbey KH, Kligman AM: The acute effects of long-wave ultraviolet radiation on human skin. J Invest Dermatol 72:253–256, 1979.

12. Kaidbey KH, Grove GL, Kligman AM: The influence of longwave ultraviolet radiation on sunburn cell production by UVB. J Invest Dermatol 73:243–245, 1979.

13. Arbabi L, Gange RW, Parrish JA: Recovery of skin from a single suberythemal dose of ultraviolet radiation. J Invest Dermatol 81:78, 1983.

14. Parrish JA, Zaynoun S, Anderson RR: Cumulative effects of repeated subthreshold doses of ultraviolet radiation. J Invest Dermatol 76:356, 1981.

15. Kaidbey KH, Kligman AM: Cumulative effects from repeated exposure to ultraviolet radiation. J Invest Dermatol 76:352, 1981.

16. Rosario R, Mark GJ, Parrish JA, Mihm MC Jr: Histological changes produced in skin by equally erythemogenic doses of UV-A, UV-B, UV-C and UV-A with psoralens. Br J Dermatol 101:299–308, 1979.

17. Smith JG Jr, Lansing AI: Distribution of solar elastosis (senile elastosis) in the skin. J Gerontol 14:496, 1959.

18. Knox JM, Cockerall EG, Freeman RB: Etiological factors and premature aging. JAMA 1979:630, 1962.

19. Smith JG, Finlayson GR: Dermal connective tissue alterations with age and chronic sun damage. J Soc Cosmet Chem 16:527, 1965.

20. Helm F: Nodular cutaneous elastosis with cysts and comedones (Favre-Racouchot Syndrome). Arch Dermatol 84:666, 1961.

21. Montagna W, Carlisle MS: Structural changes in aging human skin. J Invest Dermatol 73:47–53, 1979.

22. Braverman IM, Fonferko E: Studies in cutaneous aging: I. The elastic fiber network. J Invest Dermatol 78:434–443, 1982.

23. Braverman IM, Fonferko E: Studies in cutaneous aging: II. The microvasculature. J Invest Dermatol 78:444–448, 1982.

24. Lavker RM: Structure alterations in exposed and unexposed aged skin. J Invest Dermatol 73:59–66, 1979.

25. Montagna W, Kligman AM, Wuepper K, Bentley JP (eds): Special issue on aging. J Invest Dermatol 73:1–131, 1979.

26. Hodgson C: Lentigo senilis. Arch Dermatol 87:197, 1963.

27. Clark WH Jr, and Mihm MC Jr: Lentigo maligna and lentigo-maligna melanoma. Am J Pathol 55:39, 1969.

28. McGovern VJ, Mihm MC Jr, Bailly C, et al: The classification of malignant melanoma and its histologic reporting. Cancer, 32:1446, 1973.

29. Urbach F, Epstein HH, Forbes PD: Ultraviolet carcinogenesis: experimental, global, and genetic aspects, in Sunlight and Man: Normal and abnormal photobiologic responses. Fitzpatrick TB, Pathak MA, et al (eds). University of Tokyo Press, Tokyo, pp 259–283, 1974.

30. Urbach F: Photocarcinogenesis, in The Science of Photomedicine. Regan JD, Parrish JA (eds). Plenum Press, New York, pp 261–292, 1982.

31. Lew RA, Sober AJ, Cook N, Marvell R, Fitzpatrick TB: Sun exposure

habits in patients with cutaneous melanoma: A case control study. Dermatol Surg Oncol 9:981–986, 1983.

32. Lew RA, Koh HK, Sober AJ: Epidemiology of cutaneous melanoma. Dermatol Clin 3:257–269, 1985.

33. Kripke ML: Immunological effects of UV radiation and their role in photocarcinogenesis, in Smith KC (ed). Photochemical and Photobiological Reviews, vol 5. Plenum Press, New York, p 257, 1980.

34. Henkind P, Friedman A: Cancer of the lids and ocular adnexa, in Cancer of the Skin. Andrade R, Gumport SL, Popkin GL, Rees TD (eds). Saunders, Philadelphia, pp 1345–1371, 1976.

35. Robins P: A dermatologist's approach to the management of skin cancer. Clin Plast Surg 7:421–432, 1980.

36. Barnes HD, Boshoff PH: Ocular lesions in patients with porphyria. Arch Dermatol 48:567–580, 1952.

37. Mascaro JM, Lecha M, Herrero C, Muriesa AM: New aspects of porphyrias. Curr Probl Dermatol 13:11–32, 1985.

38. Korting GW: The Skin and Eye. A Dermatologic Correlation of Diseases of the Periorbital Region. Saunders, Philadelphia, 1973.

39. Fox LP, Fox BJ: Diseases affecting the skin and eye: a review of opthalmic manifestations. J Am Assoc Mil Dermatol 10:60–69, 1984.

40. Epstein E: Regional Dermatology: A System of Diagnosis. Grune & Stratton Inc, Orlando, pp 48–55, 1984.

41. Bernhard JD: Facial sparing in psoriasis. Int J Dermatol 22:291–292, 1983.

Protecting the Eye from Light Damage

8

Ultraviolet-Absorbing Intraocular Lens Implants

David Miller

Introduction

As our prostheses improve they tend to mimic the natural product more and more. However, in copying the human lens, we must decide upon which age of the human lens we would like to mimic because with age the human lens changes its properties. For example, the lens of a child accommodates 15 diopters, probably transmits some ultraviolet (UV) light,[*] is crystal clear, and is thinner, smaller, and lighter in weight than that of the adult. As we age, the lens thickens, loses accommodative flexibility, yellows, fluoresces, and absorbs more UV light.[1,2] Presently, the choice is somewhat simplified for manufacturers because they have no way of giving an intraocular lens (IOL) accommodation. The key variables that they can influence are power, shape, and UV transmission.

---

[*] Since it has been shown by Lerman[2] that human lens transmits UV light from childhood until the mid-20s and that adolescents reach sexual maturity at a younger age in areas of the world where ultra-light levels are high, Dr. J.J. Weiter (Chapter 4) has suggested that UV light stimulation to the retina may play a role in controlling fertility or sexual maturity in the human. These ideas are supported by measurement of the pineal gland hormone, melatonin, in humans. For example, F. Waldhauser in Vienna has shown that in childhood nighttime levels of the hormone were 40 times more than daytime levels. After puberty, nightime levels rise only tenfold. Thus it has been suggested that light-regulated melatonin may restrain sexuality and promote sleep in childhood.

The probability that UV light may be responsible for certain retinal degenerations in the elderly (see Chapter 4) helped manufacturers to decide on the spectral transmission properties of the new IOLs. Since the IOLs were primarily being put into older adults whose retinas are more vulnerable to macular degeneration, most manufacturers have decided to produce a line of UV-absorbing implants.

Theoretically, such filters would seem to have a number of additional optical benefits. Let's look at them.

Light Scattering

Studies on animal eyes have shown that the cornea scatters shorter wavelengths more intensely than longer wavelengths. Specifically, UV and blue light incident on the cornea are splashed and scattered as they strike the lens more so than green and red light.[3] Thus, placing a UV filter in the IOL would tend to eliminate this highly scattered portion of the spectrum from striking the retina and degrading the retinal image.

False Color

The above paragraph suggested that UV light can penetrate the eye and strike the retina. But are the retinal photoreceptors built to receive and transduce UV light? Stated another way, can UV light affect our perception of the world? The answer is yes. Our photoreceptors all contain sensitivity peaks in the UV range (300–400 nm).[4] Although vestigial now, such a sensitivity was useful in the past.[†]

[†] About 550 million years ago (Phanerozoic era), the first chlorophyll-containing organisms, relatives of the blue-green algae of today, produced oxygen by photosynthesis. It is estimated that their oxygen contribution to the atmosphere was 1% of today's level. This oxygen produced the early ozone layer. This earliest ozone shield was able to stop small amounts of UV light from striking the earth. About 400 million years ago the concentration of oxygen rose to about 10% of today's level, and the ozone layer began to cut out enough UV light to allow life to climb out of the sea and colonize the shore. Such life probably took the form of green plants. Then about 100 million years ago, angiosperms, true seed-bearing plants, appeared. Initially spread by the vagaries of the wind, they soon developed the attractive colors and fragrances needed to entice insects to pollinate the plants. Although the ozone layer had thickened by then,[5] longer UV radiation could still rain down upon the

Figure 8.1. Photograph on left shows the black-eyed Susan as we see it in normal light. Photograph on right is the same plant taken through special UV-transmitting lens. A pollinating insect, which sees the UV light, is given a black bull's eye to aim at. (Courtesy of Dr. Thomas Eisner.)

It seems ironic that UV light is seen again when an adult has a cataract removed. The phenomenon is then known as "false color." Figures 8.2 and 8.3 are a dramatic representation of such a phenomenon. The painter of both scenes is the same person.[7] Figure 8.2 top was painted using his phakic eye, which had a visual acuity of 20/20. Figure 8.2 bottom was painted using his aphakic eye, which can also see clearly. One might postulate that UV stimulation of the aphakic eye's retinal receptors, particularly the blue cones, does two things. Under very bright outdoor conditions, the phe-

---

earth. At that time, the insect-plant mutual dependence relationship started to mature. For example, the insect would climb into a plant's flower seeking to feed on the energy-high nectar in exchange for collecting pollen and spreading it to other plants. In such a system the plant had to advertise its presence, as well as guide the insects through its pollen-laden parts to the nectar reward at the base of the flower. With much of the ambient light being ultraviolet, the important flower markings had to reflect UV light.[6] Therefore, the early insect and bird photoreceptors had to have complimentary sensitivity peaks in the UV. Thus, the colors that we primates see in flowers might be considered incidental. A flower may look orange because its pigment reflects UV light and, incidentally, also reflects orange (Fig. 8.1). On the other hand, the UV sensitivity in our photoreceptors may have had its origins in those early insect or bird receptors. This all seems a bit wasteful because during adult life our photoreceptors aren't permitted to see UV light because of crystalline lens absorption.[2]

Figure 8.2. Original scene painted by artist using his phakic eye (top). Painted copy of original scene (bottom) as perceived by aphakic eye. (From Spitalny, LA, Devue JB, Fenske AD: Color perception in unilateral aphakia. Arch Ophth 82:592–595, 1969. (Copyright 1969, American Medical Association.)

nomenon tends to "bleach out" color appreciation in the blue, giving everything an opposing red glow (erythropsia). Under moderate lighting conditions, UV radiation stimulates each receptor in an unfamiliar manner, primarily overstimulating the blue receptor or the blue sensitivity of the other receptors, thus making reds more maroon and green more green-blue.[8] A similar phenomenon occurs in photography. Figure 8.3 shows two photographs of same scene. The photograph with the UV filter (left) permits

Figure 8.3. A similar phenomenon occurs in these photographs. The photograph on the left was taken with a UV filter while the photograph on the right was not. (Courtesy of Dr. R. Stegmann.)

less bleaching of the film and allows a purer rendition of the subtle coloring.

Chromatic Aberration

Isaac Newton[9] was the first to observe the presence of chromatic aberration in the human eye. Longitudinal chromatic aberration has been measured to be between 0.75 to 2.75 D[10] in the adult human eye, and it forms the basis of the duochrome test in the subjective refraction examination of a patient. By expanding the spectrum of wavelength that can be received by the retina in the aphakic eye, the larger the blur produced by chromatic aberration. Wald measured the chromatic aberration in the aphakic eye[11] (Fig. 8.4). This greater smear in focus could account for the perception of a faint glow around each object seen and might be responsible for the "glare" noted by some aphakic patients.

UV-Absorbing IOLs

Let's stop for a moment and look at the key factors underpinning the decision to add a UV absorber to an IOL: (1) Ultraviolet radiation (less than 400 nm) comprises about 5% of the total solar radiation striking the earth at present and less than 0.5% of the radiation from artificial fluorescent lamps.[12] (2) Laboratory experiments and clinical studies have shown that UV and blue light, in high doses, can cause retinal damage in certain animals and patients.[13–15] (3) the UV absorbing quality of the crystalline lens decreases chro-

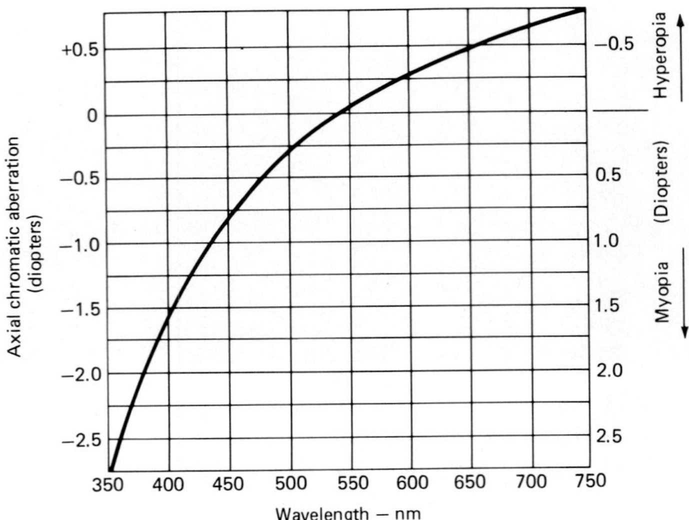

Figure 8.4. Chromatic aberration of the aphakic adult eye. (From Wald G: Human vision and the spectrum. Science 101:653–658, 29 June 1945. Copyright 1945 by the AASS.)

matic aberration, false color, and prevents some scattered light from reaching the retina. (4) One study [16] showed that the incidence of cystoid macular is more common in aphakic patients without UV-absorbing IOLs than in those with UV-absorbing IOLs. (5) Some commonly used medication—i.e., chlorothiazides, chlorpromazines, phenothiazines, protriptylines, tetracyclines, and chlorpropamides—can cause phototoxicity or photoallergy of the skin triggered by the UV spectrum. It is possible that a similar response can occur in the aging posterior segment of the eye if UV light is allowed to penetrate the IOL. (6) Cataract surgery and IOL implantation are now done with a greater frequency on younger patients. This group may be more apt to participate in skiing or other outdoor sports and thus be exposed to high levels of outdoor UV light. Theoretically, this group could be vulnerable to UV- or blue-light-induced retinal damage. Recent work suggests that there is a possibility that levels of free radical scavangers may decrease in the retina and vitreous of some aphakic patients. [17] Such changes would make this group more vulnerable to UV damage. I believe it was this line of reasoning that led to the development of the UV-absorbing IOL. Is there a chemical that manufacturers can add to an IOL which will absorb UV light safely?

Nature of UV Absorbers

Luckily, we in ophthalmology are not the first group faced with developing a product that prevents UV light damage. For years industry has contended with the fact that outdoor exposure to UV fades colors, yellows paper, stiffens rubber, weakens textiles, cracks outdoor paint, and makes plastics brittle. Actually, the manufacture of UV absorbers for plastics is a sizeable business. For example, in 1980 more than 4 million pounds of UV "stabilizers" (absorbers) were sold in the United States at an average cost of $10 per pound.[18] Ultraviolet absorbers have routinely been used in such familiar plastics as polymethylmethacrylate (PMMA), polypropylene, polystyrene, polyethylene, polyamides such as nylon, and polyvinylchloride for about 20 years.[18] It might be interesting to see how these clear plastics are damaged by UV light. For such damage to occur, the plastics must contain photosensitizing agents—i.e., substances that absorb the energy of UV radiation and then funnel such energy into the propagation of free radicals, which can then chop up the polymeric links. Such agents might be specks of catalyst left after polymerization, unwanted aromatic impurities inadvertently added at some step, residual monomer that was never properly mixed into the polymer, or new chemicals that arose during the thermal processing of the plastic. Upon UV exposure, these agents become photo excited and usually react with available oxygen to produce peroxy radicals. These radicals attach to hydrogen atoms from adjacent polymer chains, forming hydroperoxides. Such peroxides digest the polymeric links and ultimately yellow the plastic as well as produce visible cracks in the plastic.[19] Fortunately, the plastic industry's large base of experience has helped those making IOLs. For example, a number of effective nontoxic UV chromophores were already available when "ophthalmology" saw the need to incorporate UV filters into the IOL.

Aside from low toxicity, the chromophore, which must absorb UV radiation, cannot act as a photosensitizer itself, i.e., start to break down and change. One of the first UV stabilizers that came out of research in the 1950s was the family of compounds known as the benzophenones (Fig. 8.5). The two benzene rings in this family, with their alternating double bonds, absorb the long UV radiation (300–400 nm). Strategically placed hydroxyl and methoxyl groups help with further UV absorption and upon stimulation fold into a reversible six-sided ring that dissipates the absorbed energy in a harmless manner (temperature elevation of less than one degree). Because of these additional chemical groups attached to

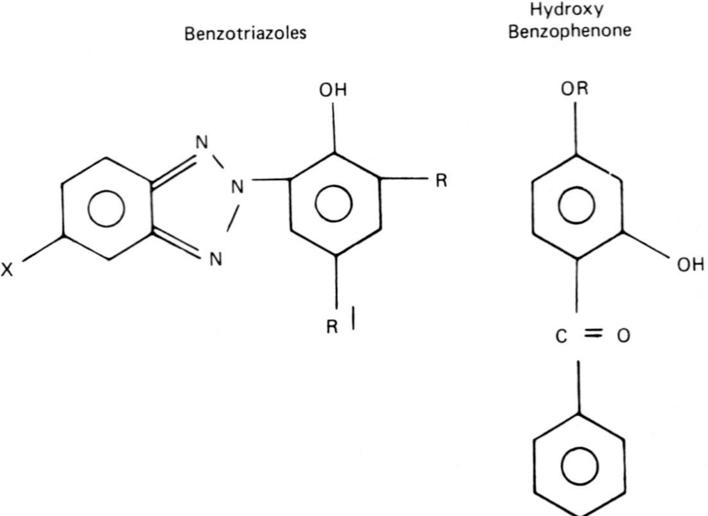

Figure 8.5. Chemical structure of a typical benzophenone and benzotria-zole.

the benzene rings, the benzophenones can be chemically linked to PMMA during its polymerization. This creates a polymer with UV stabilization built into the actual skeleton of the molecule. At least two major IOL manufacturers (at the time of this writing) produce their implants using this technique.[20]

A second competitive UV absorber that also emerged in the 1950s came from a family of compounds called the benzotriazoles (Fig. 8.5). These molecules contain two benzene rings separated by three nitrogen atoms linked to one of the benzene rings by two double bonds. One can only assume that these last mentioned double bonds allow the benzotriazoles to absorb UV light in a manner similar to the benzophenones, but at substantially lower concentrations (about $\frac{1}{10}$ to $\frac{1}{50}$ of that of benzophenones). The benzotriazole molecule also uses UV absorption to form new reversible rings, and therefore it uses the new energy in a harmless manner, never creating free radicals. Benzotriazole is not chemically bonded to PMMA; it is mixed with the plastic during the molten state, being trapped in the interstices of the polymeric chains as the plastic cools.[21]

As research in this area progresses, physiologic chromophores with useful properties might emerge, closely resembling the crystalline lens. For example, certain manufacturers now contemplate mixing a special blend of these compounds in an IOL to absorb

UV as well as some of the short blue wavelengths.[22]

One IOL manufacturer gives the IOL UV-absorbing properties during the sterilization process. The manufacturer has found that using high-energy ionizing radiation for sterilization also converts some of the components of the plastic into molecules that absorb certain properties of the UV spectrum. As yet, these components have not been identified.[23]

Loop Materials

The function of an IOL loop is primarily structural, not optical. Its job is to hold the optical component concentric with the pupil for the lifetime of the patient. As mentioned earlier, plastics can become brittle and crack when exposed to sunlight. The presence of a UV stabilizer in the plastic can retard such deterioration. Of course, only the loops of anterior chamber and pupillary plane lenses are exposed to light. These loops are currently made of polypropylene or PMMA. Blue polypropylene contains a dye that absorbs UV radiation, probably converting incident UV energy into harmless heat. Clear polypropylene is usually supplied to the IOL manufacturer with a UV stabilizer. However, this material is then heated and molded into specific loop shapes. At the time of this writing, the potency of the stabilizer is not known after such shaping and heating treatment. PMMA loops are also vulnerable to UV damage. In general, extruded PMMA loops do not contain a stabilizer, whereas one-piece IOLs will have a stabilizer/absorber if the optic has a stabilizer. The question of UV destruction of loop materials may actually be only academic, since very few cases of structrual loop failure have been reported after the implantation of well over a million IOL implants.[24]

Amount of UV Absorbers

The average dark sunglass used by a weekend sailor absorbs about 75% of the incident visible light. The "sunglass" visors used by the astronauts in space absorb 98% of the incident visible light. Obviously, the greater amount of sunlight in space requires the astronaut to wear a denser filter. Thus, the amount of light striking the retina depends on ambient light intensity as well as on the absorption of the filters used.

How much UV light should an IOL absorb? Should a patient

living in a sunny part of the world have more UV absorption than one living in the north? Perhaps the properties of the human adult lens suggest an answer. Lerman[2] found that the adult lens from age 25 to 82 absorbs an average of 80% of the UV light below 400 nm.

There is another fact of nature that may help guide us. Native people living in the tropics have more choroidal pigment to protect their retinas. Thus, the argument may be summarized by saying that an IOL that absorbs on the average 80% of the UV radiation below 400 nm best mimics the human condition. For aphakic patients living in extremely sunny environments, either IOLs of greater UV-absorbing capacity or the added use of sunglasses may be needed for extra necessary protection.

Theoretical Dangers of Chromophores in IOLs

We in medicine must always be aware of the potential dangers of additives, preservatives, artificial-coloring compounds, and in the case of IOLs, chromophore additives and PMMA impurities. Because there are so many variables involved in the manufacture of acceptable (i.e., pure) PMMA itself, we must rely primarily upon empirical tests. That is, we should use the PMMA that has the best and longest record of purity. In the case of chromophores, we have to rely upon leaching and extraction tests done under physiologic conditions. Thus, the extraction of a chromophore from an IOL by a strong solvent is not as relevant as finding the chromophore leaching out into a physiologic solution. Thus far, neither the benzophenones nor the benzotriazoles have been noted to leach out of IOLs placed in physiologic environments.[25,26] Naturally, the dangers involved in chromophore leaching would not be an issue if physiologic agents that functioned as stable chromophores could be found.

The use of the YAG laser has presented a new aspect to the potential dangers of the IOL. Some people have likened the blast of a clinical YAG laser to that of a miniature thermonuclear explosion. Remember, the temperature at the focus of the YAG is about 15,000°C.[27] Theoretically, such a concentrated amount of energy could crack polymer chains, releasing harmful monomers of PMMA, as well as agitate chromophores and drive them to the surface. Thus far, early studies have shown that IOLs with chromophores bound chemically to the PMMA polymer do not release toxic products when struck by a YAG beam at low evergy levels.[28] However, multiple bursts of the YAG laser at energy levels of 10 mJ or greater may release compounds that cause death to tissue culture cells.[29–31]

UV-Absorbing Implants: Summary

Some of the arguments for preventing UV light (below 400 nm) from reaching the adult retina seem quite convincing. They include a) histologic evidence of UV-induced retinal damage in animals, b) potential danger due to increased usage of photosensitizing drugs in older patients, c) greater numbers of younger aphakic patients spending more time in the sun, d) early evidence that there may be a lower concentration of free radical scavengers in the vicinity of the retinas of aphakic patients, and e) UV-induced erythropsia, false color, and greater chromatic aberration. The question is then, should we place the UV absorber in the IOL or prescribe UV-absorbing spectacles for all aphakic patients?

Questionable aspects of placing UV chromophores in an IOL include a) the possibility of toxic product release over long periods of time or release when the implant is struck by a high-energy laser beam, b) the increased cost of these implants and, c) inadequate knowledge of optimal amount and type of chromophore. For example, differently manufactured IOLs absorb differing amounts of UV light between 380 to 400 nm. Since we do not know how damaging or useful these wavelengths are, we cannot intelligently evaluate the manufacturer's claims. To complicate things further, we know even less about the dangers of short wavelength visible blue light in the aphakic eye. Finally, there is the relationship of UV absorption and IOL power. Higher-powered implants are thicker, and thicker implants contain more UV chromophore molecules. Thus, in comparing UV absorption between different manufacturers, one must be sure that the implants being compared are of the same thickness and power.

Reflected Light and the IOL

Thus far we have discussed the UV light focused, i.e., transmitted by the IOL. In this section we discuss the light reflected by the IOL.

Figure 8.6 is a photograph of a patient's eye with an IOL. Note the strong corneal reflection of the flash lamp as well as the obvious reflection from the IOL. All wavelengths are reflected equally from the front surface of an IOL (whether it has a UV absorber or not). If the angle of incidence is chosen carefully, the IOL can reflect a large amount of light, as seen in Figure 8.7.

However, different effects may take place in the case of anterior chamber lenses and pupillary plane lenses. The edges of these lenses are exposed to peripheral light rays. The edges being flat

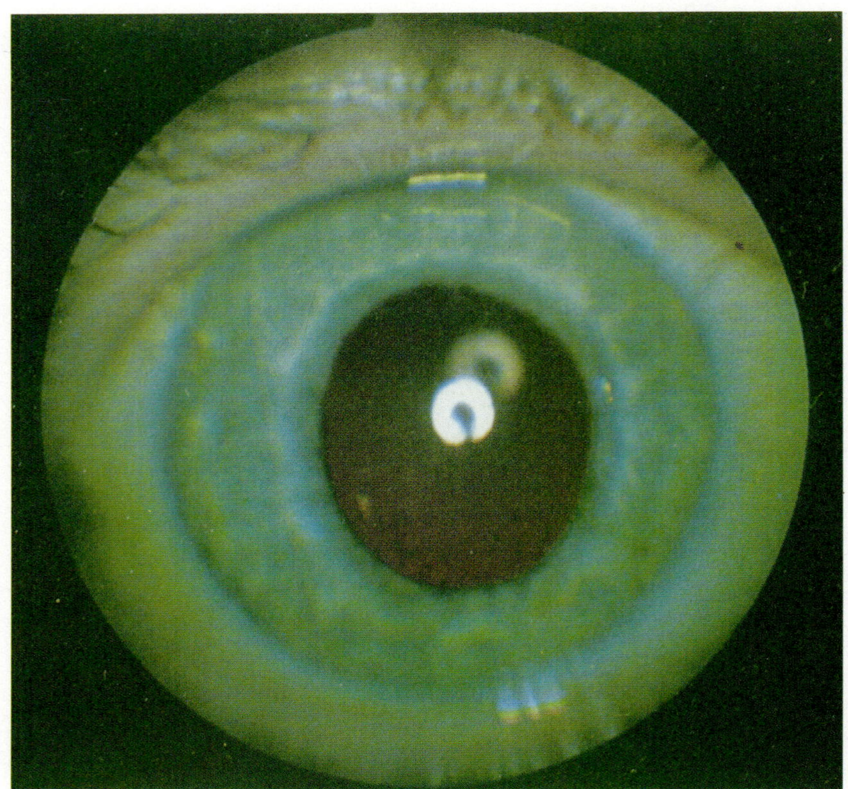

Figure 8.6. Photograph of an eye with an IOL. Note stronger corneal reflection along with significant reflection from the IOL.

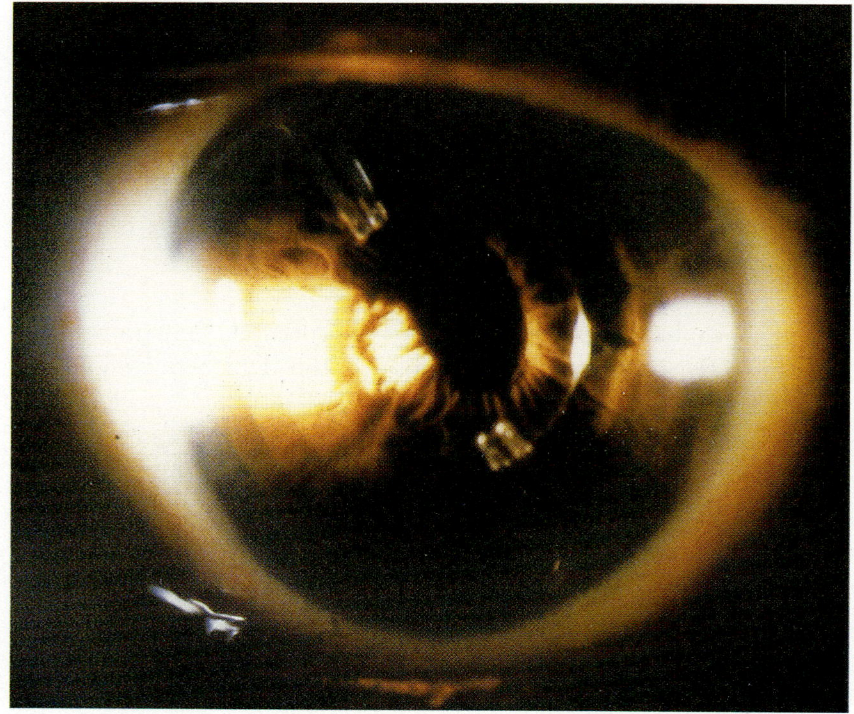

Figure 8.7. Flat edge of anterior chamber IOL which reflects light.

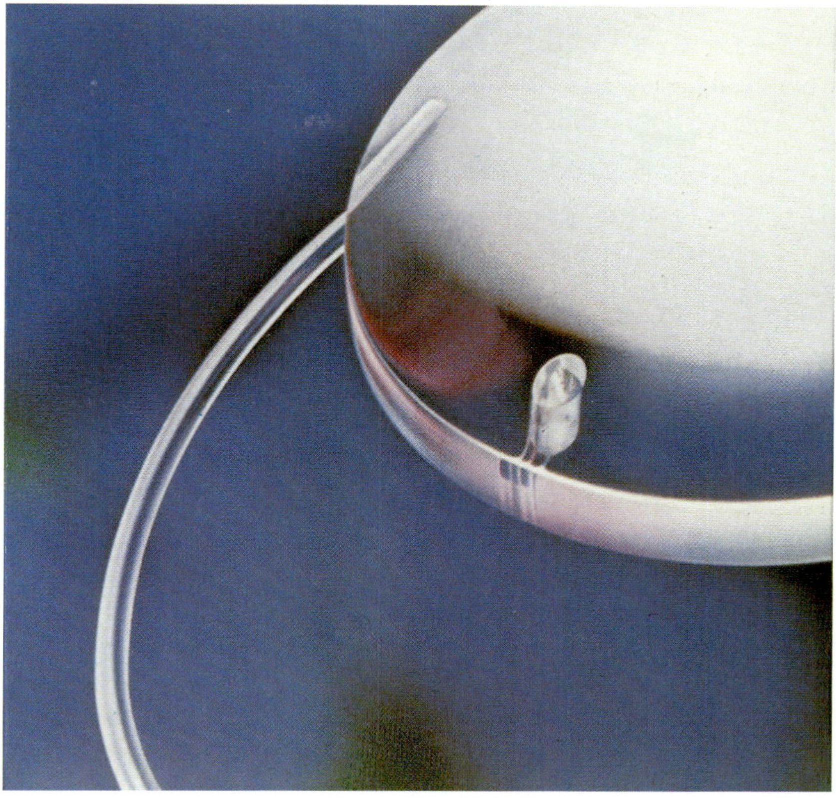

Figure 8.8. Incident beam on left is strongly reflected off the IOL. The reflected beam is seen because it is scattered by the cornea.

can either reflect light (Fig. 8.8), accept light and transmit it to the other side of the implant as a light guide, or back-scatter the incident light if the edge is roughened in any way.

Although these phenomena are all theoretically possible, one may ask if they occur at a significant level. Figure 8.9 is a clinical photograph showing a narrow slit lamp beam captured by an anterior chamber IOL, as if it were a light guide lighting up the opposite side of the IOL. In Figure 8.10 a narrow beam of light (right) is carried to the other side of the IOL and is thrown directly into the angle (note the glow at the limbus). To investigate the phenomena still further, we suspended an anterior chamber lens in a tank of slightly murky water. We then directed a narrow laser beam toward the lens edge from many different angles (Figs. 8.11, 8.12). Note that the IOL acts as a light guide and throws the incident beam either backward (to the retina) or sideway (to the angle, iris, or cornea).

Such phenomena may account for a number of clinical findings. On occasion, patients with anterior chamber lenses or pupillary plane lenses will report seeing a partially glowing arc or circle. From the results of the above experiments, we may conclude that when peripheral light strikes the flattened edge of these

159

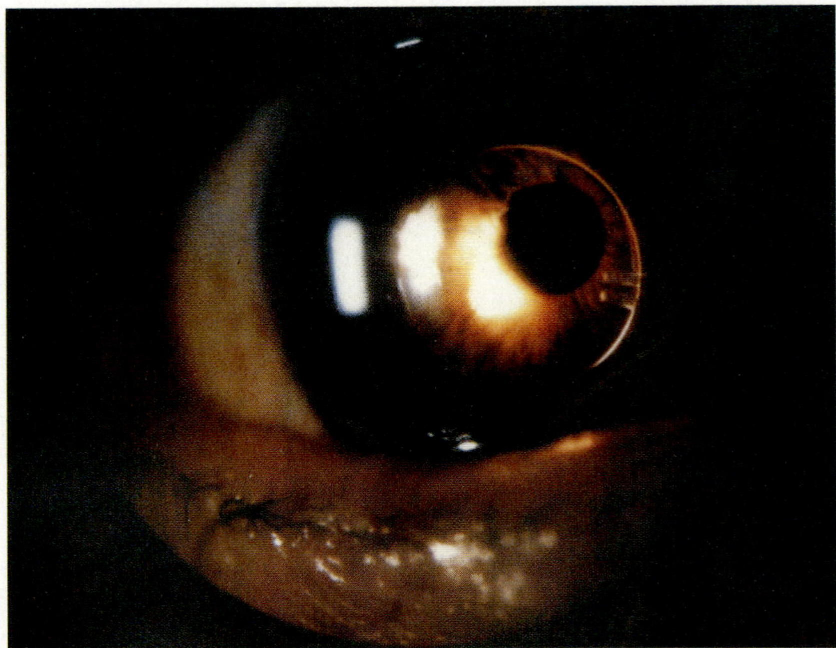

Figure 8.9. Intraocular lens accepts side illumination, then acts as a light guide carrying light to opposite side of IOL.

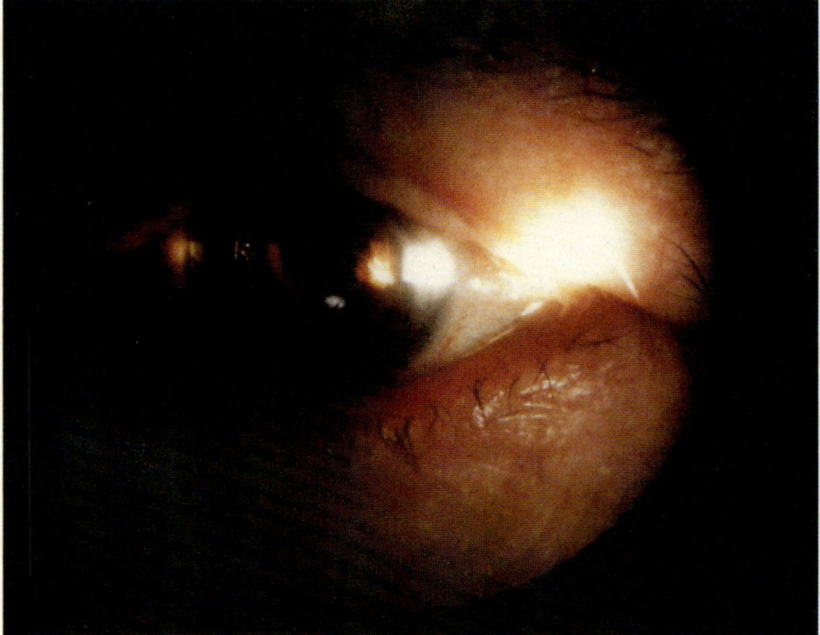

Figure 8.10. Narrow beam of light carried by the IOL and thrown into the angle on the opposite side.

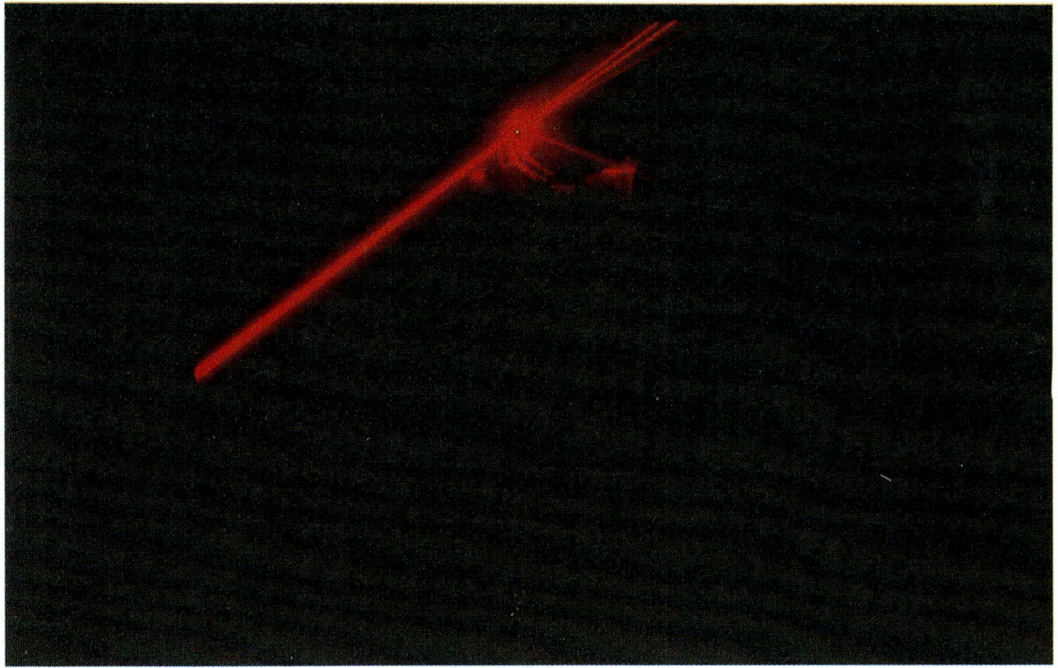

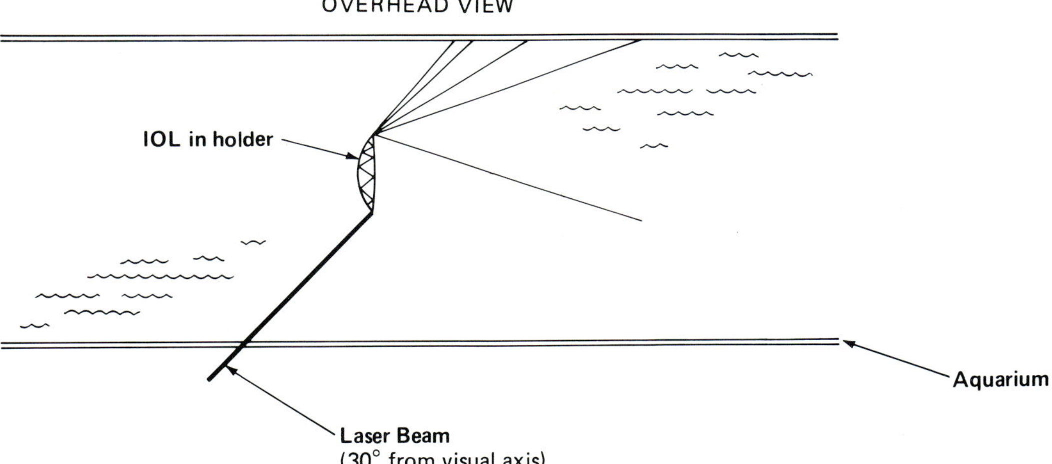

Figure 8.11. Incident laser beam (30° angle with IOL optical axis) being dispersed by IOL to opposite side.

lenses, the light is transmitted to the opposite side of the implant. Some of this transmitted light is then thrown posteriorly toward the retina. If a significant portion of the lens edge is illuminated in this way, then the patient would report seeing a lighted arc.

Another clinical finding that might be related to these experiments is the higher incidence of synechiae surrounding the inferior loops of flexible anterior chamber lenses. These lenses are gener-

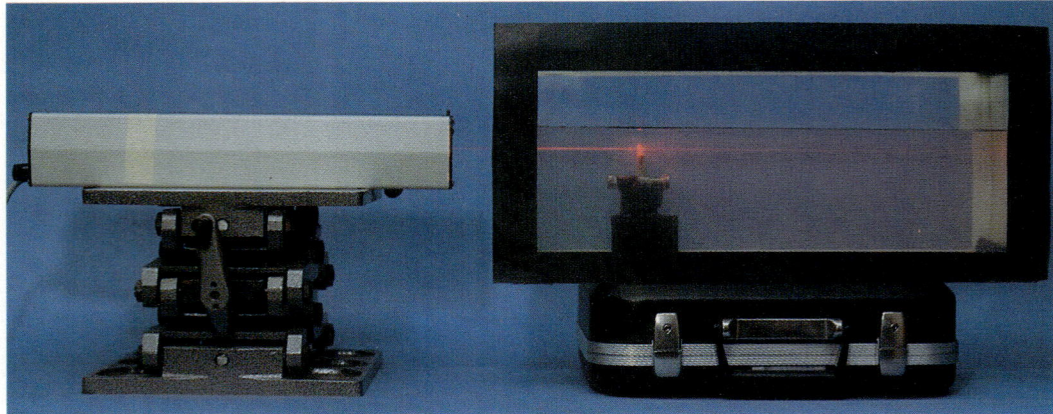

Figure 8.12. Experimental setup used for Figure 8.11. The IOL is suspended in the aquarium containing mildly murky waters.

ally implanted with the loops at the 6 and 12 o'clock positions. Since most peripheral light strikes our eyes from above or from the sides, one would expect that more implant-guided light would be directed to 6 o'clock rather than 12 o'clock. This would be in keeping with the finding that was noted in our clinic, namely, the presence of more or heavier synechiae surrounding the 6 o'clock loops as opposed to the 12 o'clock loops.

Finally, we would like to draw attention to a new idea. Normally, the angle structures of the eye do not receive light. Only through the use of a gonio lens can we artificially allow light to enter and exit from the angle. However, the presence of an anterior chamber lens or an iris plane lens can guide light (UV and visible) into the angle structures. We have learned that the presence of light, oxygen, and a suitable photosensitizer can produce free radicals, such as the superoxide radical. Such radicals can cause tissue damage and incite inflammation. The aphakic anterior chamber has a higher oxygen content than that of the phakic eye because of the absence of the oxygen-consuming crystalline lens. Thus, the presence of an anterior chamber lens in the aphakic eye provides higher levels of light and oxygen to the angle. All that is needed for free radical formation is the proper photosensitizer accumulating in the angle. Many systemic drugs taken today can be photosensitizing or break down to photosensitizers. In Chapter 4 on angle structures we were appraised of the fact that many free radical-scavenging systems are present in the angle. Thus, the possible production of free-radical-induced inflammation in the angle is dependent on light levels, oxygen levels, presence of photosensitizers, and the level of opposing scavenger systems. We submit that in some patients the levels of the above can be so skewed that high levels of free radicals can develop and result in inflammation[32] in the interior of the eye. Common photosensitizing drugs include diazide diuretics used in hypertension, oral hypoglycemics such as chlorpropamide or tolbutamide, antibiotics such as tetracyclines, chloramphenicol and certain sulfa drugs, and tranquilizers such as the phenothiazines.[33] Thus, the presence of any of these drugs in the anterior chamber or angle in the proper concentration could help trigger an inflammatory reaction in patients with anterior chamber or pupillary plane IOLs. Hopefully, these thoughts will stimulate further study in this area.

References

1. Weekers Y, DelMarcille J, Luyehx-Bacus J, Collignon J: Morphological changes of the lens with age and cataract, in The Human Lens in Relation to Cataract. Ciba Foundation Symposium, Elsevier, Amsterdam, pp 25–41, 1973.
2. Lerman S: Radiant Energy and the Eye. Macmillan Publishing Inc, New York, 1980.
3. Farrell RA, McCally RL, Tatham PER: Wavelength dependence of light scattering in the normal and cold swollen rabbit corneas and their structural implications. J Physiol 233:589–612, 1973.
4. Schichi H, Dratz EA (eds): Proceedings: Symposium on molecular aspects of the visual process. Photochem Photobiol 29:655–745, 1979.
5. Young LB: Earth's Aura. Alfred A Knopf, New York, pp 50–63, 1977.
6. Lythgoe JW: The Ecology of Vision. Chardon Press, Oxford, 1979.
7. Spitalny LA, Devue JB, Fenske AD: Color perception in unilateral aphakia. Arch Ophthalmol 82:592–595, 1969.
8. Tan K: Vision in the ultraviolet. Doctoral Thesis, Utrecht, Holland, Rijksuniner slteit de Utrecht, 1971.
9. Newton I: Opticks, ed 4, Book I, Part 2, Prop. VIII (Reprinted by Bell, London), p 165, 1931.
10. Mandelbaum T, Sivak JG: Longitudinal chromatic aberration of the vertebrate eye. Vis Res 23(12):1555–1559, 1983.
11. Wald G: Human vision and the spectrum. Science 101:653–658, 1945.
12. "Terrestrial-Global Spectral Irradiance Tables for Air Mass 1.5," ASTM Document 138 RI, E44.02, Feb. 1981.
13. Ham WT, Mueller HA, Sliney D: Retinal sensitivity from short wavelength light. Nature 260:153–155, 1976.
14. Ham WT, Ruffolo JJ, Mueller HA, et al: The nature of retinal radiation damage. Dependence and wavelength, power level and exposure time. Vis Res 20:1105–1111, 1980.
15. Ham WT, Ruffolo JJ, Mueller HA, et al: Histological analysis of photochemical lesions produced in rhesus retina by short wavelength light. Invest Ophthalmol 17:1029–1035, 1978.

16. Kraff M, Sanders DR, Jampol LM, Lieberman H: Effect of an ultra-violet-filtering intraocular lens on cystoid macular edema. Ophthalmology 92:366–370, 1985.
17. Jampol LM: UV-light toxicity and its management. Ocular Surgery News Symposium. Ocular Surgery News, Slack Publishing, Thorofare, NJ, p 6, 1985.
18. Hardy WB: Commercial aspects of polymer photostabilization, in Allan NS (ed): Developments in Polymer Photochemistry, vol 3, Applied Science Publishers, London, pp 287–346, 1982.
19. Abovelezz M, Water PF: Studies on the photodegradation of poly-(methylmethacrylate) PB 297793, June, U.S. Dept. Commerce NBS. National Technical Information Service, Washington DC, 1979.
20. Paulson DR: A report on chemically bonding UV absorbing molecules to PMMA. Optical Radiation Corporation, Azusa, CA, May 1984.
21. Goldberg EP: UV absorbing intraocular lenses. Presented at American Intraocular Implant Society, Los Angeles, 1984.
22. Pharmacia Ophthalmics, Uppsala, Sweden: Personal Communication, 1984.
23. Zigman S: Tinting of intraocular lens implants. Arch Ophthalmol 100:998, 1982.
24. Hessberg PC: Prolene versus PMMA: The controversy continues. Ophthalmology Times. June 15, pp 3,16,17, 1983.
25. Toxicity Testing. Intermedics Intraocular Inc, Pasadena, CA, 1984.
26. Toxicity Testing. Optical Radiation Laboratories, Azusa, CA, 1984.
27. Barnes PA, Rieckhoff, KE: Laser induced underwater sparks. Appl Phys Lett 13:282, 1968.
28. Stability of UV Bloc™ Lens. IOLAB Corp, Covina, CA, 1984.
29. Terry AO, Stark WJ, Newsome DA, Maumenee AE, Pina E: Tissue toxicity of laser damaged intraocular lens implants. Ophthalmology 92:414, 1985.
30. Lerman S: Observations on the use of high power lasers in ophthalmology. IEEEJ. Quantum Electronics. QE.20:1465, 1985.
31. Lindstrom RL, Mowbray SL, Skelnik S: Neodynium: YAG laser interaction with intraocular lenses: An in vitro toxicity assay. Am Intraoc Implant Soc J 11:558–564, 1985.
32. McCord JM: Free radicals and inflammation. Protection of synnovial fluid by super oxide dismutase. Science 185:529–531, 1974.
33. Haber LC, Kochevar IE, Shalita AR: Mechanisms of photosensitization to drugs in humans, in The Science of Photomedicine, Regan JD, Parrish JA (eds). Plenum Press, New York, pp 323–349, 1982.

9

Approaches to Protection Against Light-Induced Changes in the Eye

David Miller and Robert Stegmann

Over the course of evolution, nature has developed a number of ways of protecting the eye and the body from the harmful effects of light. In this chapter the principles of these natural methods of protection are discussed, followed by a description of man's adaptation of these principles to augment his own protection. However, before we launch into discussion of how to prevent light from striking the eye, we should look at sunlight itself.

When and Where Ultraviolet Light Changes

The combined light of sky and sun, known as global illumination, is not a constant entity. The longer the distance that light travels in a scattering media (i.e., our atmosphere), the smaller the amount of short wavelength radiation (ultraviolet) that will arrive on the earth's surface. Thus, the farther one is from the earth's equator, the longer the penetration distance of sunlight and the lower the amount of arriving ultraviolet (UV) light. A similar argument affects the incident light at different times of the day. As our earth rotates, we come closest to the sun at about noon, and that's when we get our greatest share of UV light. As the day lengthens, so does the distance from the sun. By late afternoon, not only has all of the UV light been scattered, but the short blue wavelengths have also been lost by scattering and the sky takes on a red glow. Using a similar argument, the elliptical orbit of the earth around the sun means that during certain seasons we are farther from the sun and thus the light penetration distances are longer, the exposure to our scattering atmosphere greater, and therefore the amount of incident UV light is less.[1a]

Finally, two other environmental factors will influence the amount of UV radiation that will strike our eyes. Because clouds are made of small water droplets, they scatter light differently than the very fine particles that make up our atmosphere. Whereas particles of a size smaller than the wavelengths of visible light preferentially scatter shorter wavelengths (Rayleigh scattering), large particles such as water droplets scatter all wavelengths in a similar manner (Mei scattering). Thus UV light will penetrate the clouds to the same extent as visible light.

The last factor to be noted is the variation of UV-reflecting properties of our environment. Although our northern climes receive less UV radiation, the snow-covered surfaces will reflect from 60% to 85% of the incident light into the eyes of the inhabitants. Therefore, the total amount of radiation is almost equivalent to that faced by a Nomad looking down at the Sahara desert, where the sand will reflect only 17% of the incident UV radiation. The same almost holds true for the Samoan native facing the waters of the tropical Pacific, where only 5% of the incident UV light is reflected back.

Thus, by way of summary, one might say that good UV protection would be most needed on a cloudless spring day at noontime on the ski slopes located in Spain or those of the northern Andes.

Naturally, these arguments do not apply to artificial lighting. Incandescent bulbs produce only negligible amounts of UV radiation, and bright white fluorescent lighting yields less than 1% of that found naturally on a sunny summer's day.

Broad Spectral Permanent Light Absorption

Natural Pigment

The pigment melanin is a brownish-black pigment, collected in descrete packages within cells called melanocytes. The pigment is found throughout the animal kingdom, ranging from its presence in the protective ink of the squid, to the skin of man. Interestingly, the number of melanocytes in a black skin is essentially identical to that in white skin. The amount of melanin is virtually zero in an albino skin and is slightly less than 1 g in the body of an adult black person.[1] The "heart" of the chemical structure of the melanin molecule is the amino acid tyrosine. The enzyme tyrosinase controls the major steps in the biosynthetic pathway whereby tyrosine is oxidized through dopaquinone to be converted to the protein polymer melanin. Melanin absorbs all wavelengths from UV

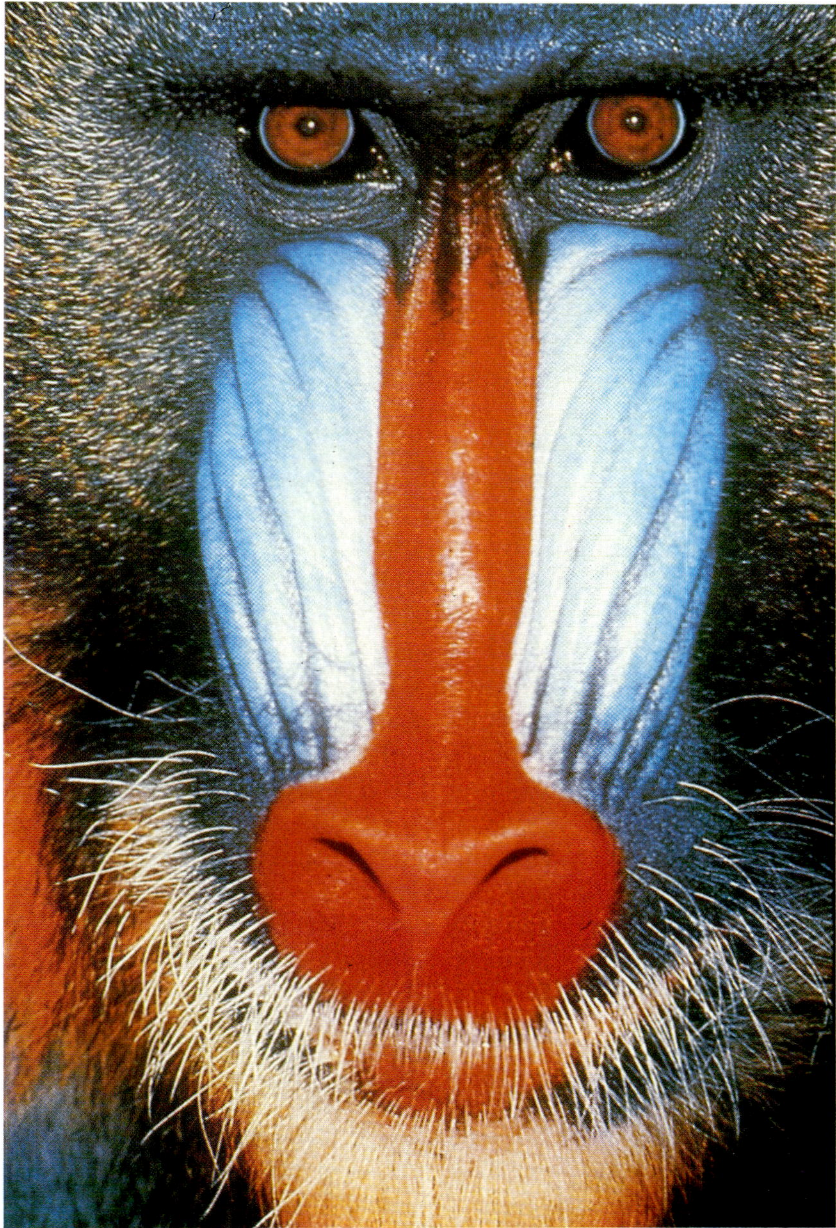

Figure 9.1. The Mandrill monkey. Note the darkly pigmented sclera. (Courtesy of Günter Ziesler.)

to the far red. In the eye, melanin is primarily found in the pigment epithelium of the iris, choroid, and retina. Thus, melanin is used to control the amount of excess light from overwhelming the lens, vitreous, and retina. Figure 9.1 shows it also to be present in the sclera of certain monkeys (Mandrill monkey). One would assume that this creature is exposed to levels of light high enough to significantly penetrate the sclera.

Sunglasses

Man often uses this "across the board" light absorption approach with dark sunglasses. The starting elements of a sunglass are either clear spectacle glass (crown glass), which absorbs some UV-A, UV-B, and UV-C (i.e., below 350 nm); the plastic polymethyl-methacrylate (PMMA), which absorbs almost no UV light; or the plastic polycarbonate, which absorbs the light below 380 nm. Dyes are then added for cosmetic appearance and extra light-absorbing qualities.

The important variables in sunglasses are the spectral absorption qualities of the dye and the concentration of the dye. These parameters have been shown by Segre et al[2] to vary considerably from one commercial sunglass to another. In their study of 50 randomly selected commercial sunglasses in Italy in 1981, they found that UV absorption in the 300 to 400 nm range (a) varied from almost none to 100% and (b) varied from absorbing all the wavelengths in the 300 to 400 nm range to only portions within this range. On the other hand, Magnante et al[3] showed that nine different, randomly selected, inexpensive plastic clip-on sunglasses purchased in the United States in 1985 all filtered out more than 85% of the long UV rays, i.e., 300 to 400 nm (Fig. 9.2). As noted by Lerman in the last chapter of this book, only a few brands filter out 99% of these wavelengths.

As a general rule, most strongly tinted yellow, green, gray, and brown sunglasses effectively filter out UV light. Generally blue-tinted glasses are poor UV absorbers.

Other Sunglass Features

Wide Temples. Since bothersome light can enter the eye from the side, some sunglasses have protective side shields. These can be permanent or in the form of detached plastic or leather pieces, or simply a widened side piece (known as the temple of the frame). Figure 9.3 shows the removable wraparound. Although useful when sailing or skiiing, they can reduce side vision and are out-lawed for driving in some states.

Visors, Flip-Ons. Since the main source of glare for most outdoors-men is the sun, some way of blocking light from above makes vision sharper and more comfortable. Heavily constructed frames offer this function as do flip-on sunglasses, which can be set in visor position or flipped down to cover the eyes.

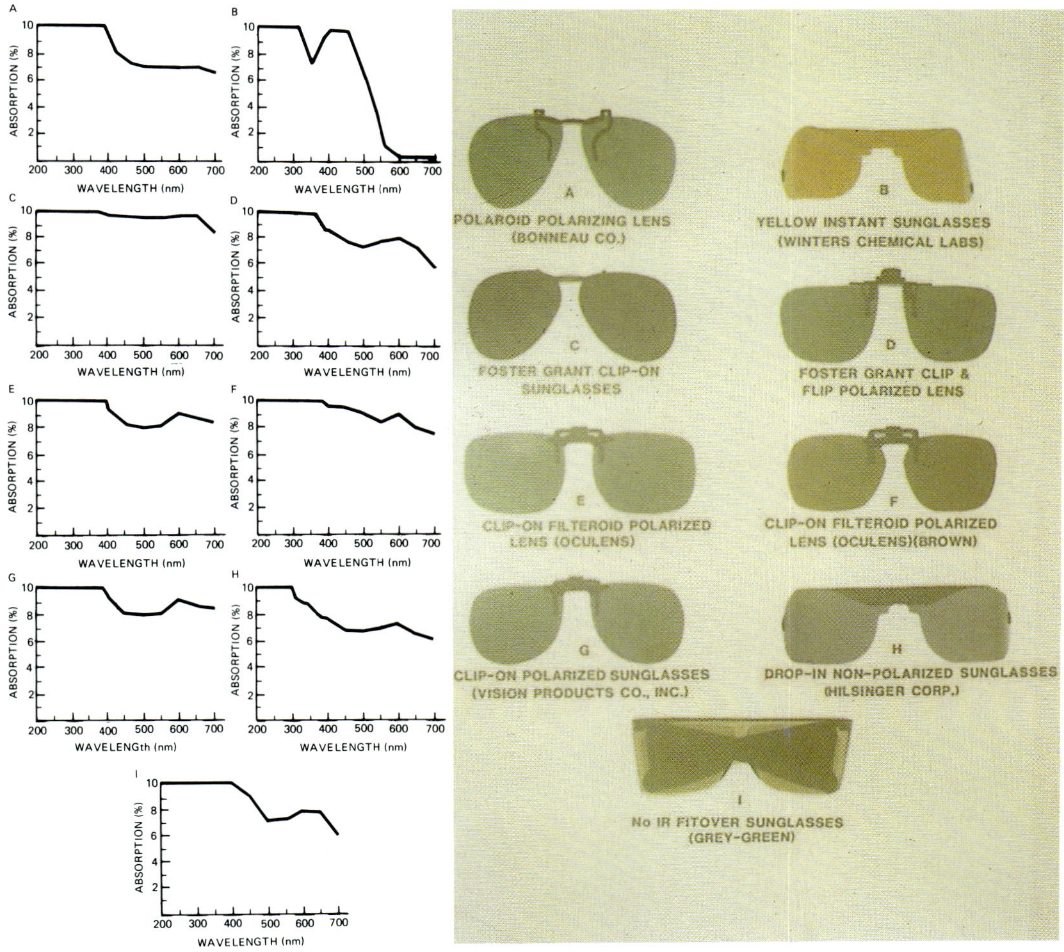

Figure 9.2. The light absorption characteristics of nine inexpensive, plastic clip-on sunglasses purchased in the United States. (From Magnante D, Miller D: Ultraviolet absorption of commonly used clip-on sunglasses. Annals Ophth 17:614–616, 1985.)

Gradient Glass. These glasses are dark at the top and progressively lighten toward the bottom. Their function, much as the visor, is to eliminate annoying light from above. These glasses can be safely worn at night or indoors (Fig. 9.4).

Frames. The sunglass lens provides the all important absorbing qualities and determines the basic price. It is the design and construction of the frame that provide the sturdiness as well as the glamour and style. These last items produce the wide range in pricing found in the marketplace. For example, stronger, nonblemishable hinges with locking screws similar to those seen in prescription spectacle frames are found in more expensive sunglasses. Plas-

Figure 9.3. Wraparound attachment on regular spectacles.

tic sidepieces (temples) are more expensive if reinforced with a metal rod. Metal frames made of precious metal are more expensive than those stamped out of plastic.

Broad Spectrum Temporary Light Absorption

The Pupil

When the skin is exposed to UV light, new melanin is synthesized, and a tan develops within a few days. The eye of the animal also responds to light, but in a faster mode. The pupil of the eye

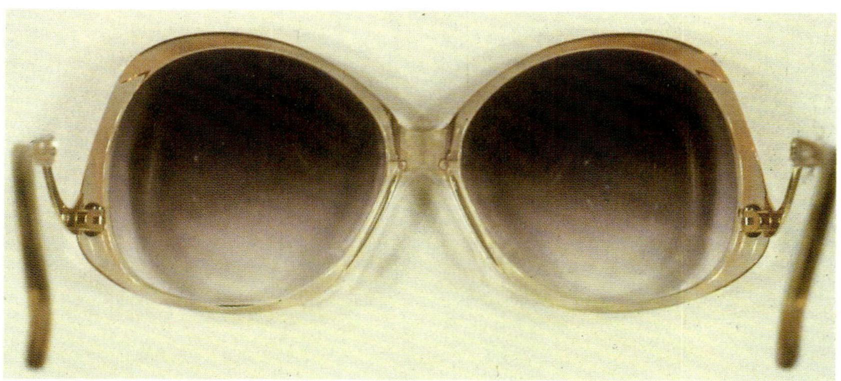

Figure 9.4. Graded-density sunglasses.

constricts as the area of the iris expands in less than a second after light exposure. Thus, the pupil can decrease the amount of light reaching the retina and lens by a factor as great as 16 in less than 1 second.

By attenuating the light striking the eye, do sunglasses induce dilation of the pupil and so partially cancel the effect of the sunglasses? Keeping in mind the many factors involved in measuring pupil size (light duration, light amplitude, spectral distribution, pupil adaptation, fatigue, environment variation in light reflectance, etc), let us refer to the data provided by the Illuminating Engineering Society[5a] and look at what happens on a very bright day.

The irradiance on a bright day will range from 10,000 to 30,000 foot-lamberts. A dark pair of sunglasses will reduce light by 80%. Thus, the retinal-pupil system will receive 2,000 to 6,000 foot-lamberts, which is still about ten times brighter than a brightly lighted indoor environment. Note that this all occurs where the curve plateaus (Fig. 9.5). Thus, under the sunglass attenuated light, the pupil may dilate about 0.25 mm. This could maximally translate to an increase in pupil area of about 10%. This would mean an increase of 10% in UV light incident upon the lens and retina if the sunglass provided no UV-absorbing capability. In fact, just about all sunglasses manufactured today will absorb at least 75% of the incident UV radiation. Therefore, we can conclude that the pupillary dilation behind most sunglasses would allow from 0% to about 3% more UV light to strike the lens and retina.

Photochromatic Sunglasses

After learning that differently shaded glasses help in different situations, one naturally wonders if a sunglass could be made to get

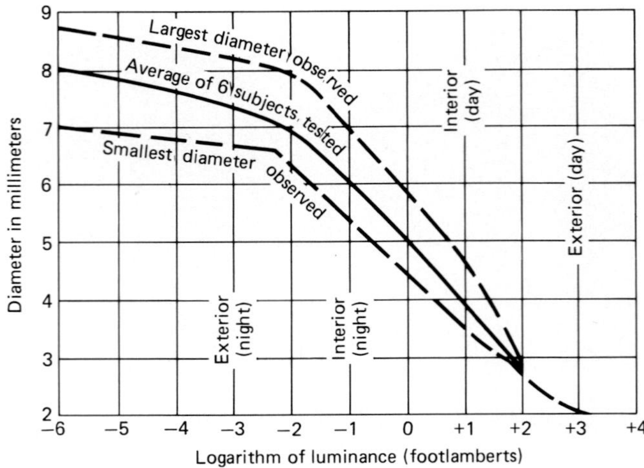

Figure 9.5. The effect on pupil diameter when the eyes are exposed to a large, uniform field of fixed luminance. (From Reeves P: Response of average pupil to various intensities of light. J Opt Soc Am 4:135, January 1920.)

darker as the surroundings get brighter and lighter as the environment gets dimmer. After all, automatic cameras respond to different levels of illumination with their built-in light meters and computerized circuitry. In fact, photochromic glasses accomplish the same thing, but without the circuitry. Buried within the glass are silver halide crystals much as they are found in the emulsion of a piece of film. When light between 300 and 400 nm strikes a single fleck of silver halide on the film, it permanently turns the fleck dark as it changes to pure silver. That same reaction takes place in the glass, but with one difference. The change is not permanent. Thus, glass impregnated with these crystals can change shades forever in response to the surrounding light levels.[3] The original photochromic glasses came in two versions. Photogrey could go from 15% absorption indoors to 55% out of doors, whereas photosun went from 35% absorption indoors to 80% in sunlight. Newer versions, known as ultraphotochromic, have a wider range of response, going from 20% absorption indoors to 80% out of doors. These lenses take one or two minutes to darken and about 20 minutes to lighten when one comes in from the sunny outdoors. These lenses strongly absorb UV light in their darkened state.

Figure 9.6. Deep-sea fish with yellow pigmented corneas.

Partial Light Absorption

Nature's Pigmented Corneas

Figure 9.6 shows an example of yellow-pigmented corneas in a deep-sea fish. These pigments primarily absorb UV and blue light. In the visual world of these fish, such wavelengths produce confusion either by being preferentially scattered by particles in water, or the atmosphere provides an annoying background when the fish look up. Therefore, the yellow filters help cut down glare as well as make certain contrasts sharper for the fish.

The human also develops natural protective pigmentation under certain special conditions. Figure 9.7 is an example of a corneal scar infiltrated with melanin pigmentation. In each case, a patient with a heavily pigmented brown iris sustained a traumatic perforation associated with an iris prolapse. Although all lacerations were repaired surgically and each iris reposited back into the globe, iris pigment remained at the wound site. In time the pigmentation almost appears woven into the healing scar. Thus, by absorbing a significant amount of incident light, which would be scattered by the scar, the pigmentation helps reduce glare sensitivity.

Experimental Alteration of the Corneal Absorption Spectrum

Since skin sunscreens have proven effective in eliminating the unpleasant effects of UV-induced sunburn, we wondered if a similar approach might also be useful for eye protection. Such a sunscreen would have to be nontoxic to the eye, be retained in the cornea for a useful period of time, and absorb the UV light below 400 nm. As a first step we turned to the ophthalmic formulary and asked, "Was there a drug already approved for eye use with the optical properties that we wanted?" After a careful search we noted that the antibiotic tetracycline had the UV-absorbing properties needed. In a series of experiments we observed that the application of 1% tetracycline ophthalmic ointment to rabbit corneas resulted in a significant increase in the amount of UV light absorbed by the cornea.[5]

Of course, tetracycline itself is not the ideal ocular sunscreen. Continual human use could result in the development of strains of bacteria resistant to the antibiotic. It could also upset the normal flora balance in the conjunctival sac. Finally, tetracyclines are photosensitizing agents and might actually produce damage to the ocular tissue over time. Nevertheless, this first foray into the area of ocular sunscreen has shown that the idea is feasible.

Altering Light Transmission via Light Scattering

The Sclera

The sclera is a tissue that only allows about 1% of the incident light to penetrate the eye.[6] The sclera is opaque to light by the mechanism of light scattering. The most common example of light scattering in nature is the white-topped waves of the ocean. These white caps are composed of ocean water mixed with air bubbles. Although both water and air are themselves transparent, each has a different index of refraction, i.e., a different ability to bend light. When light is caught in a mixture of two materials with different indices of light, and these materials are widely separated physically (i.e., more than the size of a wavelength of light), and are randomly arranged, then little light gets through. The result is that most of the light is tossed and turned and ultimately thrown backward. In the sclera, large randomly placed collagen fibers of one index of refraction are separated from each other by a mucopolysaccharide ground substance of another index of refraction. Thus, the sclera appears white because of backscattering of most

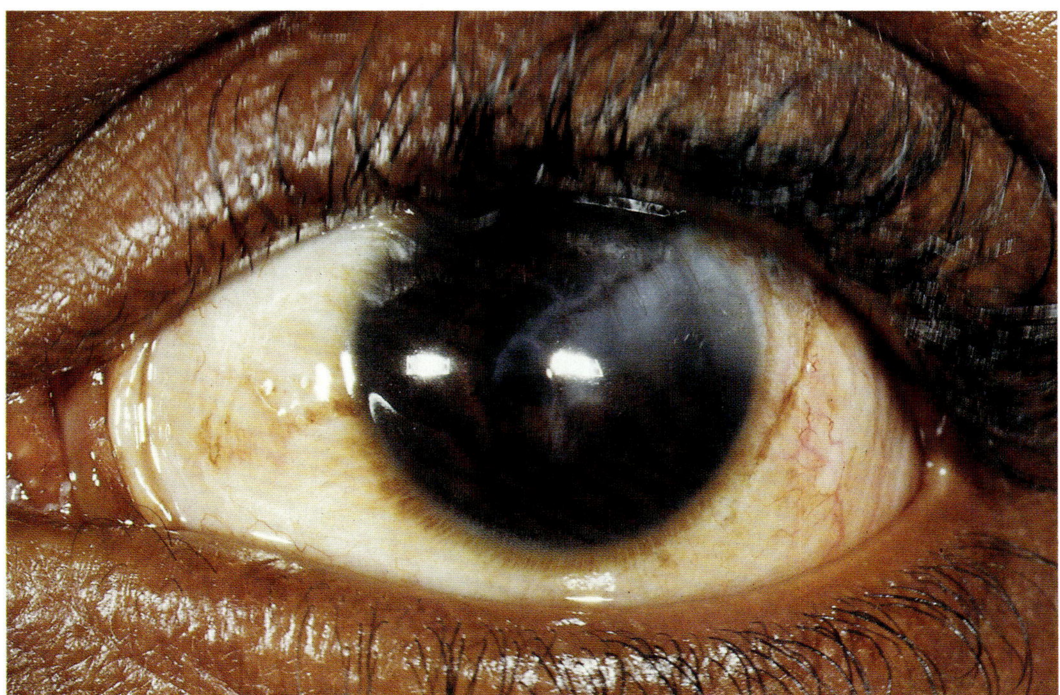

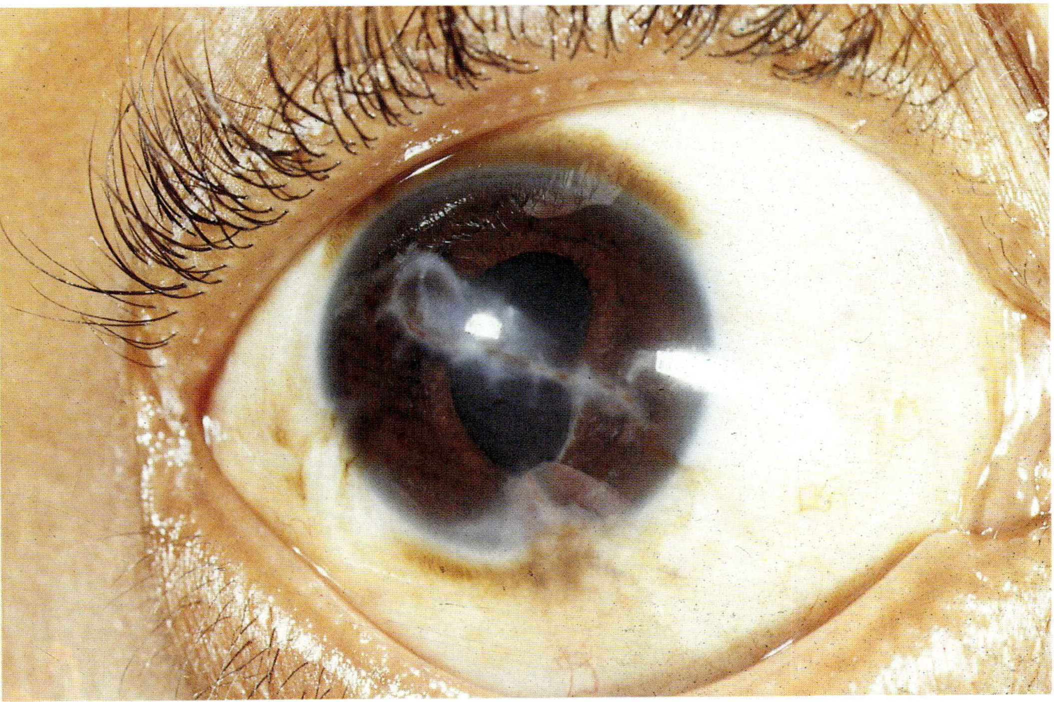

Figure 9.7. Two patients with heavily pigmented irises who sustained corneal lacerations associated with a iris prolapse. Note the pigmentation that has been "woven" into the healing scars.

of the incident light.[7] When the scattering elements are larger than a few hundred nanometers, all wavelengths are scattered in a similar fashion.

Man's Use of Light Scattering

For people with light-sensitive skin, sunblocking agents are often recommended. These agents are white creams. In general, they are made of particles of either titanium dioxide, talc ($MgSiO_2$), bentonite (FeO_2), or zinc oxide in an ointment base. As in the ocean's white caps, the powder, particles, and base have different indices of refraction and thus backscattering of the light occurs, never allowing it to reach the skin.

Man has commonly used the principle of light scattering to modulate the light in his home. Lace curtains and woven lamp shades are made up of mixtures of fibers and air spaces that scatter light. Translucent shower screens and bathroom window glass also scatter light but in a different manner. Their irregular surfaces can be thought of as arrangements of glass lumps and air valleys that scatter light much as the white caps of the sea.

Nature's Use of Reflection and Refraction

In the eye there are two regions in which very little light is allowed to enter. One is the area just behind the iris,[10a] i.e., the ciliary body and the ora serrata.[5]

The other protected area is the angle of the eye. It is the shape of the cornea that prevents light from reaching the angle. The average anterior chamber angle subtends a solid angle of about 30°. Thus, although the cornea can be thought of as accepting a solid angle of 180° of light, the anatomic angle of the eye can theoretically only accept a narrow cone from this source. To clarify this point, supposing you were sitting in the angle of the eye, looking out. Your view would be a bit cramped by the overhanging scleral shelf above and the iris below. Light directed toward the angle (say the nasal three o'clock side) must impinge on the midperiphery of the cornea of the opposite side (the temporal nine o'clock side). Because of the steep curve of the cornea in this area, all the light directed to the angle is bent into the iris root, never reaching the angle. On the other hand, light striking the limbus, headed toward the angle by skimming along the direction of the iris plane, is reflected away because the flatter peripheral cornea provides a "more glancing" surface—that is, a surface that provides an angle of incidence that exceeds the critical angle for these glancing rays. Therefore, these rays are totally reflected and never reach the angle. The gonio lens or gonio prism defeats the corneal optics by neutralizing the curvature.

The Koeppe gonio lens provides a new, artificial, very steep and constant corneal curve that allows rays almost parallel to the iris to enter and then leave the angle. The gonio prism, on the other hand, cancels the curvature of the cornea with its flat outermost surface. Thus, light directed toward the angle is bent, but a little, and can reach the angle. The emerging light then strikes positioned mirrors in the gonio prism and gives the observer convenient views of the angle.

Man's Use of Reflection

By covering glass with metallic particles, a mirror is made. Since the metallic particles of any single layer do not perfectly interdigitate, spaces are left for light to penetrate. Thus, a thinly mirrored surface will reflect most incident light, but allow some light to penetrate. Commonly used mirrored sunglasses take advantage of this thin layering and probably prevent about 75% of incident light of all wavelengths from reaching the eye. This would be equivalent to a darkly shaded sunglass or a photochromic glass in its darkest state. Astronauts in space are exposed to very high light levels. Therefore, they use more heavily coated gold-mirrored visors, which prevent 98% of the light from reaching their eyes.

Free Radical Scavenger

Nature's Use of Pigment

As noted in Chapter 2, in 1900 a medical student by the name of Raab noted that when the dye acridine orange was added to an aerated culture of paramecia, which was exposed to light, the paramecia died.[8] It was later learned that in the presence of light and oxygen, the dye produced free radicals that were lethal to the organism. The process became known as photodynamic action.[8] In 1957 Sistron et al[9] reported an important related phenomenon. They were studying the physiology of the purple sulfur bacteria, *Rhodopseudomonas spheroides*, an organism that contained chlorophyll and carried on photosynthesis. In the course of the work, a mutant blue-green strain that was isolated stopped growing and died in the presence of light and air. The mystery of the death of the mutants was solved when it was learned that chlorophyll generated free radicals in the presence of light and oxygen. The purple strain was protected by the pigment carotene, which is a free-radical scavenger. However, the mutant strain contained no such protection against photosensitization. Work that followed suggested that carotenoid pigments in all photosynthetic organisms (bacteria, algae, and higher plants) play an important role in protecting these organisms against the seriously damaging effects of photo-oxidation by their own endogenous photosensitizer, chlorophyll.[10]

Man's Use of Carotene

In 1961 a group of investigators at the New York University School of Medicine took the lead from Sistron's group and demonstrated that the onset of erythema induced by artificial UV light could be delayed by oral administration of beta carotene.[11] Thus, the stage was set for the use of beta carotene in patients with a serious photosensitive disease. The disease first chosen was erythropoietic protoporphyria. In this disorder, abnormally elevated levels of protoporphyrin IX in the blood produce a skin sensitivity to light, manifested by edema, erythema, and a burning sensation. Of the 133 patients originally studied, 84% developed an increased tolerance to sunlight when they took oral beta carotene. Further work has shown that beta carotene may be of help in other skin diseases such as congenital porphyria and polymorphous light eruption.[11] Work on animals has also suggested that beta carotene, along with two related carotenoids, canthaxanthin and phytoene, help to prevent the development of UV-induced skin tumors in mice.[12] As one reviews this work, it seems very possible that the use of the right oral photosensitizing inhibitors may also be of use in the prevention of light-induced eye and lid-skin disease.

Conclusion

This chapter has attempted to show how nature has evolved a number of ways of protecting the skin and eyes of many creatures from light damage (pigments, light scattering, reflection, free radical scavengers). The chapter has then tried to show how man has also used similar principles in designing protective aids against light damage.

References

1. Pathak MA, et al: Sunlight and melanin pigmentation, in Smith KC (ed): Photochemical and Photobiological Reviews. Plenum Press, New York, pp 211–239, 1976.

1a. Thorington L: Spectral, irradiance, and temporal aspects of natural and artificial light in The Medical and Biological Effects of Light, Wurtman RJ, Baum MJ, Potts JT (eds). Ann NY Acad Sci vol 453, pp 28–55, 1985.

2. Segre G, Reccia R, Pignalosa B, Pappalardo G: The efficiency of ordinary sunglasses as a protection from ultraviolet radiation. Ophthalmic Res 13:180–187, 1981.

3. Magnante D, Miller D: Ultraviolet absorption of commonly used clip-on sunglasses. Ann Ophthalmol 17:614–616, 1985.

4. Araujo RJ: Photochromic glass, in Treatise on Material Science and Technology, vol 12, Glass 1: Interaction with Electromagnetic Radiation. Academic Press, New York, 1977.

5. Krah DB, Saragas S, Miller D: Induced corneal absorbance of ultraviolet light. Ann Ophthalmol 16:818–822, 1984.

5a. IES Lighting Handbook. Kaufman JE (ed): Illum Engineering Society. New York, pp 2–3, 1966.

6. Spillman L: Density, light scatter and spectral transmission of a scarred human cornea. Albrecht Von Graefe Arch Klin Exp Ophthalmol 184:278–286, 1972.

7. Miller D, Benedek G: Intraocular Light Scattering. CC Thomas, Springfield, IL, 1973.

8. Blum HF: Photodynamic Action and Disease Caused by Light. Reinhold, New York, 1941.

9. Sistron WR, Griffiths M, Stanier RY: Biology of a photosynthetic bacterium which lacks colored carotenoids. J Cell Comp Physiol 48:473–515, 1956.

10. Krinsky NI: Carotenoid protection against oxidation. Pure Appl Chem (51):649–660, 1979.

10a. Nelson LB, et al: Aniridia, A Review. Surv Ophthalmol 28:621–642, 1984.

11. Mathews-Roth MM: Beta-carotene therapy for erythropoietic protoporphyria and other photosensitivity diseases in The Science of Photomedicine, (ed.) Regan JD, Parrish JA (eds). Plenum Press, New York, pp 409–440, 1982.

12. Mathews-Roth MM: Anti tumor activity of Beta Carotene, Canthaxanthin, and Phytoene. Oncology 39:33–37, 1982.

Overview of Light Damage to the Eye

10

Light-Induced Changes in Ocular Tissues

Sidney Lerman

Introduction

The eye is the only organ or tissue in the body (aside from the skin) that is particularly sensitive to the non-ionizing wavelengths of optical radiation (wavelengths longer than 280 nm) normally present in our environment. In addition to infrared and visible radiation, we are constantly exposed to ultraviolet (UV) radiation (solar and man-made) throughout life.

Since the normal cornea, aqueous, ocular lens, and vitreous are almost completely transparent to all the wavelengths of visible light (although the aging lens does absorb increasing amounts of shortwave visible radiation), one would not anticipate photodamage to these tissues from visible radiation. Only the retina appears to be susceptible to photodamage from visible radiation, particularly the shorter wavelengths affecting the blue cones. Furthermore, for non-ionizing radiation to exert an effect, it must be absorbed. Whereas the retina contains specific compounds (chromophores) whose function is to absorb visible radiation (photoreceptor rods and cones and macular pigment), the other ocular tissues anterior to the retina have very few chromophores that can absorb these wavelengths of visible radiation. Nature has provided us with transparent ocular media that are essentially avascular and contain very few visible wavelength-absorbing (400–700 nm) chromophores. Thus they can effectively transmit as well as refract the specific wavelengths of light required to initiate the visual process by photochemical reactions. However, these transparent tissues do have the ability to absorb varying amounts of

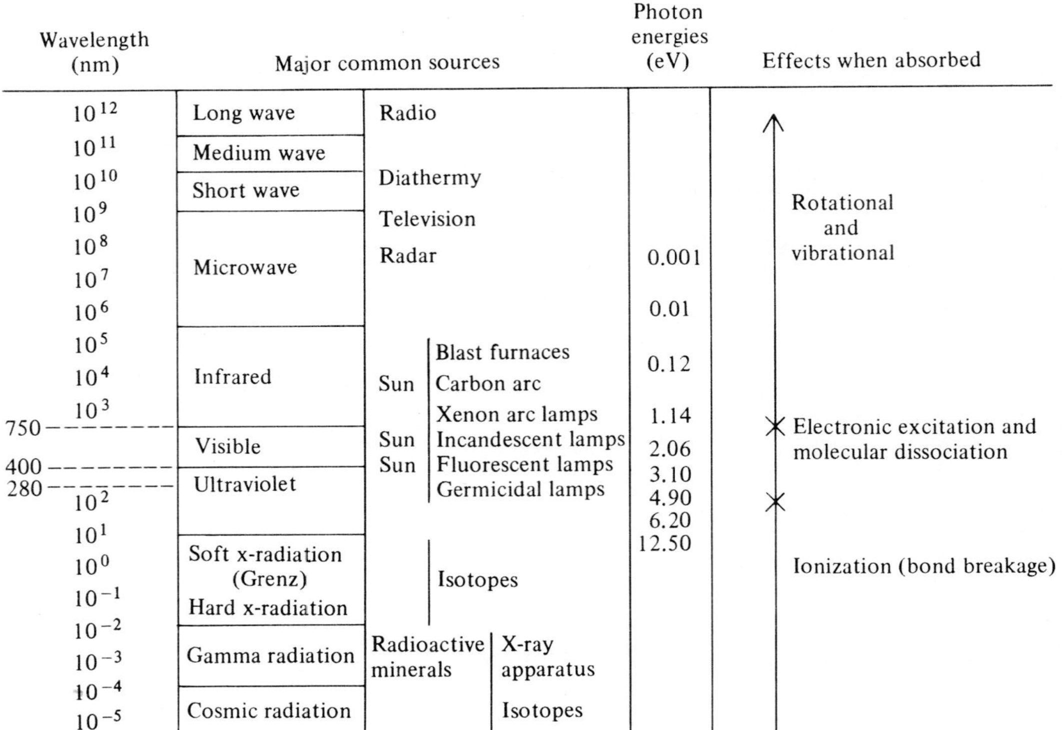

Wavelength (nm)	Major common sources			Photon energies (eV)	Effects when absorbed
10^{12}	Long wave	Radio			Rotational and vibrational
10^{11}	Medium wave				
10^{10}	Short wave	Diathermy			
10^{9}		Television			
10^{8}	Microwave	Radar		0.001	
10^{7}					
10^{6}				0.01	
10^{5}			Blast furnaces		
10^{4}	Infrared	Sun	Carbon arc	0.12	
10^{3}			Xenon arc lamps	1.14	Electronic excitation and molecular dissociation
750	Visible	Sun	Incandescent lamps	2.06	
400		Sun	Fluorescent lamps	3.10	
280 — 10^{2}	Ultraviolet		Germicidal lamps	4.90	
10^{1}				6.20	
10^{0}	Soft x-radiation (Grenz)		Isotopes	12.50	Ionization (bond breakage)
10^{-1}	Hard x-radiation				
10^{-2}					
10^{-3}	Gamma radiation	Radioactive minerals	X-ray apparatus		
10^{-4}					
10^{-5}	Cosmic radiation		Isotopes		

Figure 10.1. The electromagnetic spectrum. (From Lerman S: *Radiant Energy and the Eye.* New York: MacMillan Publishing Company. Copyright 1980 by Sidney Lerman, M.D.)

UV radiation (280–400 nm), particularly the ocular lens. The shorter the wavelength of radiation absorbed, the greater potential for photodamage, as there is an inverse relationship between a wavelength and the photon energy associated with it (Fig. 10.1). Thus, in the electromagnetic spectrum, UV radiation is the nonionizing region that could cause the most damage, provided it is absorbed. This applies to all the ocular tissues, including the retina in the very young eye (where the lens has not as yet become an effective UV filter) and in the aphakic and pseudophakic eye. In particular, the ocular lens sustains the greatest amount of photochemical change during a lifetime of exposure to ambient UV radiation.

With respect to the intensity of UV radiation in the ambient environment, it is estimated that approximately 8% (11 mW/cm^2) of solar radiation above the atmosphere falls in the 300 to 400 nm wavelength region. At sea level this is decreased to 2 to 5 mW/cm^2, depending on geographic location and season.[1]

Ultraviolet-induced changes in human and animal ocular tissues can be attributed to two mechanisms: 1) a direct or intrinsic process in which the radiation is absorbed by specific, naturally occurring chromophores within these tissues (e.g., the nucleic acids or aromatic amino acids), and 2) an indirect or photosensitized process in which the radiation is initially absorbed by photosensitizing drugs or other extrinsic compounds that have managed to penetrate into the ocular tissues.

The normal human cornea and aqueous humor transmit almost all UV radiation longer than 300 nm, although there is a small but progressive decrease in the percentage of UV radiation transmitted by the aging cornea.[1] This may be due to an accumulation of UV-induced chromophores in the cornea as it ages. Although the aqueous humor does contain small amounts of an amino acid (tryptophan) capable of absorbing some of the UV radiation that penetrates the cornea, there is little likelihood that such exposure will significantly affect this fluid. The aqueous turns over every 1½ to 2 hours; in the normal eye it is constantly being replenished and continuously exits through the anterior chamber drainage pathways. Tryptophan photoproducts that could be formed while the aqueous is circulating within the anterior chamber could conceivably enter the lens and cause some photochemical changes in this organ. However, most of the photochemical damage occurring in the ocular lens can be attributed to direct UV radiation exposure to this organ.

Lenticular Photodamage

During the past decade a considerable amount of evidence has accumulated implicating UV radiation (300–400 nm) as a significant factor in the in vitro generation of lens pigments and in protein cross-linking associated with lens aging and cataractogenesis in the mouse, rat, and human lens.[2-24] In vivo studies have also demonstrated that UV rays (longer than 300 nm) are capable of generating experimental cataracts in mouse, rat, rabbit, and primate lenses, and human UV radiation cataracts have also been reported.[25-27] The human ocular lens is constantly exposed to ambient UV radiation (300–400 nm) throughout life. One consequence of such cumulative photochemical damage is an increasing absorption of UV radiation and visible light owing to the presence of intrinsic, photochemically generated lens chromophores (pigments), which increase in concentration and number as the lens ages. At least two such chromophores have been partially characterized.[14-16] One absorbs at 360 nm and fluoresces at 440 nm; the second absorbs

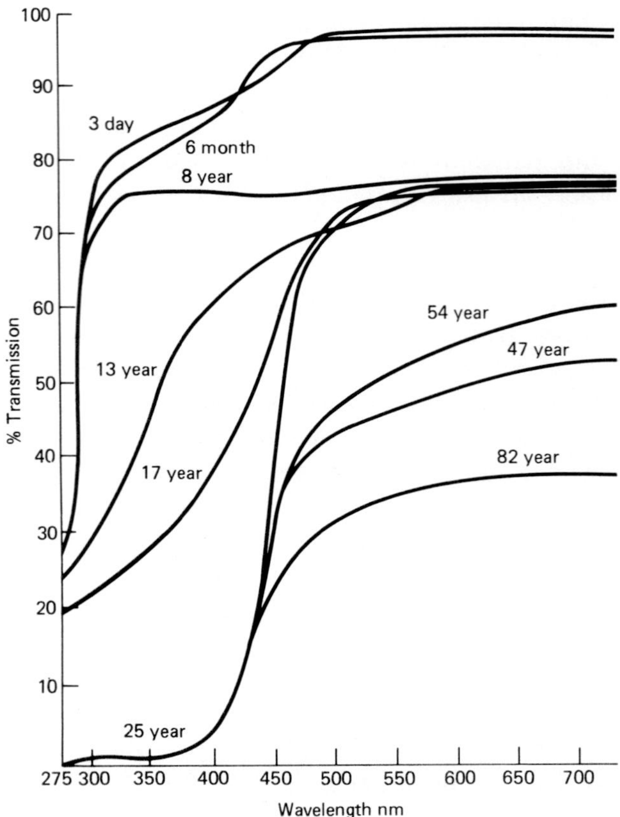

Figure 10.2. Transmission changes in the aging ocular lens.

at 420 nm and fluoresces at 520 nm. These fluorescent compounds increase in number and in concentration as the lens ages; the lens nucleus becomes yellower, and there is a progressive decrease in the transmission of visible light as well as UV radiation with age (Fig. 10.2). Discoloration is mainly confined to the lens nucleus since the cortex has much higher levels of glutathione and other compounds capable of aborting most of these photochemical reactions.[15,20] Extreme examples of this age-related photochemical generation of such lens pigments are the brown and black nuclear cataracts.

Although only a small amount of UV radiation from the sun enters the eye under normal circumstances, the cumulative effect of many years' exposure is significant, particularly when one considers man's ever increasing life span. Epidemiologic surveys provide some support for the thesis that solar UV radiation plays a role in lenticular aging and senile nuclear cataracts.[28–31] For example, cataracts, and the rate of cataract extraction, are much higher in India, Pakistan, Nepal, and certain areas of Africa than in the temperate zones. An epidemiologic investigation into the relationship between sunlight and cataract in the United States [28] reported that ". . . cataract to control ratios for persons aged 65 years or older were significantly larger in locations with large amounts of sunlight. . . ." Obviously other factors play a role in cataractogenesis including heredity, nutrition, metabolism, etc.

Thus it is now generally accepted that chronic exposure to UV radiation (300–400 nm) over an individual's lifetime leads to the generation and increased accumulation of a series of chromophores in the lens, which are responsible for the increased yellow color of the lens nucleus as it ages. In about 10% of our population this process progresses at a more rapid pace, resulting in the development of the brown (nuclear) cataract. In moderation this type of discoloration is actually beneficial since it enables the lens to become a very effective filter for UV and short wavelength visible radiation (by the second to third decade), thus protecting the retina from cumulative photochemical damage that could occur during one's lifetime. Recent studies have shown that such radiation can cause irreversible retinal photodamage in the aphakic Rhesus monkey and even in man.[32–35] It is interesting that nature has provided us with the ability to develop a lenticular UV filter to protect the retina from continuous radiation exposure that could be harmful, particularly in the older individual where the retinal metabolism and repair processes are no longer as effective as in the young. We are now beginning to see confirmation of the hypothesis that chronic exposure to long wavelength UV radiation may play an enhancing role in certain retinal diseases (e.g., cystoid macular edema, which tends to occur in older patients following removal of their cataractous lenses) and even in degenerative processes (macular degeneration and retinitis pigmentosa).

Aside from the effect of chronic exposure to ambient UV radiation, exposure to higher radiation levels can produce cortical opacities in human, rat, and rabbit lenses in vitro and in vivo. Thus, more intense UV radiation (300 nm and longer) can incude lens changes involving the cortex whereas chronic exposure mainly affects the lens nucleus. These effects appear to be dose and time related.[36] It is postulated that the more intense UV radiation is capable of inducing photochemical damage to important enzyme systems (e.g., catalase, glutathione dehydrogenase). There is also the possibility that it may affect the protein/water order within the lens,[37] resulting in the formation of "water lakes" associated with protein aggregates. This would give rise to localized areas of marked change in the refractive index in the injured site and result in light scattering and opacification.

Retinal Photodamage

Aging is known to be characterized by a loss of rod and cone cells. The recent observation that photic trauma can damage the receptors suggests a potential cumulative action of light resulting in an enhanced loss of visual cells over a period of years.[1,38–44] In other words, phototoxic effects may be cumulative in the normal aging process of the retina. Photon energies in the electromagnetic spectrum increase as the wavelength decreases, from 1.6 eV at 750 nm to 3.3 eV at 400 nm and higher energies in the UV wavelengths. These are capable of penetrating to the retina in the young eye and aphakic or pseudophakic eyes. One would anticipate that photic damage would be greatest for UV radiation (320–400 nm) and short-wave visible light in the blue region and decrease with increasing wavelengths of light, with the least damage occurring with red light. Recent studies strongly implicate long-wave UV and short-wave visible radiation (320–450 nm) as a significant factor in retinal photodamage in primates as well as other experimental animals, and even in man.[32–35,38–44] These data are of particular concern in young patients, in patients with aphakia and pseudophakia who are on photosensitizing drugs, and in all patients exposed to prolonged or above ambient levels of UV-A (315–400 nm) radiation.

The spectral sensitivity of the human retina plays a role with respect to the efficiency of a specific wavelength in producing retinal damage. The ocular lens also protects the retina from visible as well as UV radiation, since it filters more or the shorter wavelengths of visible light (blue) as compared with the longer wavelengths. Thus, the aging retina, which metabolically should be more susceptible to photodamage caused by visible light (as well as UV radiation), is in fact protected by the ocular lens, which increasingly filters out the UV and shorter wavelengths of the visible spectrum as the person ages. This might explain why human retinas are normally capable of withstanding much higher thresholds of radiation intensity as compared with other animals such as the rat, rabbit, and pigeon.[40] The retinas of these animals can be damaged by levels of environmental light that are not damaging to the normal human eye.

Photosensitized UV Radiation

A photosensitizing agent can be defined as a compund whose chemical structure endows it with the ability to absorb optical radiation (UV and visible) and to undergo a primary photochemical reaction resulting in the generation of highly reactive and relatively long-lived intermediates (triplets, radicals, and ions) that can cause chemical modifications in other (nearby) molecules of the biologic system. Regardless of subsequent events, the primary event of photosensitization in ocular tissues appears to be the absorption of light by the photosensitizer molecule, eventually resulting in a complex of photosensitizer with nucleic acids or proteins. Only the pyrimidine bases in the nucleic acids are susceptible, and in the proteins, histidine is the most readily altered of all the amino acid residues (either alone or as part of the protein) followed by tryptophan, tyrosine, cystine, and methionine. Hundreds of chemcial compounds are capable of acting as photosensitizing agents, and most of these compounds contain a tricyclic ring structure. Such compounds include a variety of drugs used in routine medical practice as well as in industry, agriculture, and the home. The most common photosensitizing reactions involve the skin, although internal organs are also susceptible to phototoxic reactions.

Only two groups of compounds (phenothiazines and psoralens) have been clearly identified as intraocular photosensitizing agents capable of causing photochemical damage to the choroid, retina, and lens in man as well as in experimental animals.[45–62] This may be due to the screening effect of the blood aqueous and the blood-retina barriers. It is well known, for example, that it is difficult, if not impossible, to obtain an effective concentration of certain antibacterial, antimycotic, or antibiotic drugs in the interior of the normal eye, and this has been attributed to the relative impermeability of the blood-aqueous and blood-retinal barriers. Furthermore, it has already been demonstrated that the cornea, and particularly the lens as it ages, provides effective filters for UV radiation and even for the shorter wavelengths of the visible spectrum, thereby nullifying any potential photosensitizing action of drugs (which are activated at these wavelengths) that might accumulate in the posterior half of the globe. However, any photosensitizing agent that accumulates in the ocular lens or retina might be a potential hazard if it becomes photobound to macromolecules within these tissues, since it would now be permanently retained there.

In addition to the demonstrated direct photochemical action of UV radiation on ocular tissues, particularly the cornea, lens, and retina, there is the possibility of photobiologic damage by means of photosensitized reactions due to the accumulation and retention of certain drugs within these tissues. For example, after the 13-mm stage of development the ocular lens is completely encapsulated and never sheds its cells throughout life. Thus, photobinding a drug to the lens proteins and nucleic acids ensures its lifelong retention within the lens, with the potential for enhanced photo-damage if these compounds are capable of acting as photosensitizing agents. A similar situation exists in the neural retina, which does not regenerate.

There are several parameters that could influence the possibility and extent of photosensitized damage in the eye. These include 1) the chemical structure of the drug in question and its absorption spectra; 2) the role of the blood-aqueous and blood-retinal barriers; 3) the age and/or health of the patient; and 4) the oxygen concentration in the affected tissue.

Certain drugs having a tricyclic (three-fused aromatic rings), heterocyclic, or porphyrin ring system are known to be efficient photosensitizers. The reason for this is 1) they have long-lived triplet states, 2) they have a low oxidation potential (type 1 reaction), and 3) their triplet state energy is such that transfer of energy to ground state oxygen is possible (type II reaction).

The psoralen compounds are well-known photosensitizing agents and have been used (under controlled conditions) in many dermatology clinics to treat psoriasis and vitiligo.[63–64] This form of phototherapy, commonly referred to as PUV-A therapy, involves the ingestion of 8-methoxypsoralen (8-MOP) or related compounds followed by exposure to UV-A radiation (320–400 nm) for short periods of time. 8-Methoxypsoralen can be found in a variety of ocular tissue within two hours after the animal (rat, dogfish, and monkey) or human is given a single dose (equivalent to a therapeutic level) and can become photobound to lens proteins and DNA if there is concurrent exposure to ambient levels of UV-A radiation.[65] Since the mature ocular lens is an effective filter for UV-A radiation in most mammals (including man), there can be no photobinding of 8-MOP in the retina. However, UV-A radiation can penetrate to the retina in aphakic and pseudophakic experimental animals and in young eyes (where the ocular lens still permits significant penetration of UV-A radiation), and 8-MOP photobinding can occur in such retinas.

PUV-A therapy and cataract formation have been documented in human cataracts as well as in experimental animals.[46–53,56,61,62] Cataracts from patients on PUV-A therapy were subjected to high-resolution phosphorescence spectrosopy.[61] The lens proteins from these patients showed phosphorescence peaks identical (in shape and lifetime) with the previously reported 8-MOP lens protein photoproducts seen in experimental rat PUV-A cataracts.[50,57] These data provide objective proof that in human lenses this drug can generate specific photoproducts that have been shown to be associated with the formation of PUV-A cataracts in experimental animals. However, this observation should not deter anyone from prescribing such therapy for psoriasis since simple and effective preventive measures are available. It should be noted that free 8-MOP can be found in the lens *for only 24 hours*, provided that the eye is protected from UV radiation exposure.[50,53,60] Thus, many dermatologists are now providing proper UV-filtering glasses to all their PUV-A patients, with instructions to put them on as soon as they ingest the drug and continue to wear them for at least 24 hours. They must be worn indoors as well as out of doors, since there is sufficient UV-A radiation in ordinary fluorescent lighting to photobind the 8-MOP. A 3-year follow-up study utilizing UV slit lamp densitography has proved the efficacy of this approach.[60,65] All the patients wore proper UV-filtering glasses for at least 24 hours following drug ingestion, and none developed enhanced or abnormal lens fluorescence levels. In contrast, patients whose eyes had not been protected (those treated prior to 1978) had anomalous and enhanced lens fluorescence and some developed PUV-A cataracts. It should be noted that PUV-A therapy could pose a potential hazard not only to the ocular lens but to the retina in young people whose lenses are not effective UV absorbers and/or in aphakic or pseudophakic individuals, particularly if they are exposed to repeated PUV-A therapy. The clear polymethylmethacrylic acid (PMMA) intraocular lenses currently in use are excellent transmitters of UV radiation [1] and thus provide less protection from UV-A radiation than the natural lens or even ordinary glass (which can absorb UV radiation up to 320 nm). Ultraviolet-absorbing intraocular lenses are now being tested by several manufacturers and should provide a simple solution for preventing potential UV-A photodamage to the pseudophakic retina.

Preliminary studies on the effect of 8-MOP and exposure to UV-A in both albino and pigmented mice, where much of the UV-A can be transmitted to the retina, demonstrate that there are several receptor organelles that can be damaged by such exposure.[66,67] These include the rod outer segments (ROS), the ellipsoid region of the rod inner segments, and the receptor nuclei. The damage involves vesiculation and disruption of the ROS, swelling of the ellipsoid with damage to the mitochondria, and necrosis of the nuclei. UV-A by itself selectively causes more damage to the ROS, whereas 8-MOP in combination with clinical UV-A radiation exposure results in far greater necrosis of the nuclei in addition to ellipsoid and ROS damage. Photobinding of 8-MOP and DNA with cross-linking upon irradiation could account for the observed necrotic process.

It should be emphasized that under most circumstances, the retina would be protected from UV-A damage by the ocular lens.

After the first decade of life, the ocular lens becomes an effective filter for all UV radiation entering the eye (300–400 nm) and even acts as a partial filter for short wavelength visible radiation (400–500 nm). However, the potential for retinal damage in children (whose lens has not yet become an effective UV-A filter) and in those with aphakia and pseudophakia has recently been evaluated. (PMMA intraocular lenses currently employed are actually excellent transmitters of UV-B as well as UV-A radiation.) Using ^{3}H and ^{14}C labeled 8-MOP, autoradiographic techniques have shown that this drug becomes photobound in young monkey retinas (<1 year of age) and in aphakic and pseudophakic monkey retinas at any age following a single PUV-A treatment.[62] Similar data had previously been reported in experiments involving other species, but the studies on primates are probably more relevant to the human condition. The UV-A radiation exposure levels used in these experiments were relatively low (never exceeding 0.4 mW/cm^2), indicating that indoor fluorescent lighting could also be hazardous with respect to retinal 8-MOP photoxicity. A similar situation has been shown to pertain to the generation of 8-MOP photoproducts in the ocular lens. These data demonstrate that patients must wear proper UV-filtering goggles for at least 24 hours following ingestion of the drug. Studies have shown that most of the unbound 8-MOP leaves the lens and retina during this period if these tissues are protected from UV-A exposure.

The phenothiazines, especially chlorpromazine, have adverse ocular side effects.[45,68,69] Chlorpromazine is reported to be cataractogenic, with pigment deposits appearing in the lens as well as in the cornea and conjunctiva. The role of NP-207 in retinopathy is well established while that of thioridazine and chlorpromazine is far less understood. There is a marked accumulation of chlorpromazine, prochlorperazine, and thioridazine in the uveal tract of pigmented animals. The phenothiazines, like other polycyclic aromatic compounds (e.g., quinolines and tetracyclines), have a high affinity for melanin. The relationship between such an affinity and functional alterations in vision is as yet unclear. With NP-207 the pigmentary pattern parallels the rod and cone disruptions. With other phenothiazines, such as thioridazine, few side effects have been reported when administered in doses of less than 800 mg/d. Chlorpromazine has been used extensively as an antipsychotic. It also binds to melanin and causes relatively minor pigmentary alterations of the fundus, although some alterations in visual functions have been noted.

Allopurinol is a commonly used antihyperuricemic agent in treating gout. Scattered reports have appeared regarding the possible relationship between the development of lens opacities in relatively young patients (second to fourth decade) and chronic ingestion of this drug.[70] Cataracts obtained from 11 patients on chronic allopurinol therapy (>2 years) were subjected to high-resolution phosphorescence spectroscopy. The characteristic allopurinol triplet was demonstrated in all the cataracts.[71] Identical spectra were obtained on normal human lenses incubated in media containing 10^{-3} molar allopurinol and exposed to 1.2 mW/cm^2 UV radiation for 16 hours; control lenses (irradiated without allopurinol) were negative. Similar data were obtained on lenses from rats given one dose of allopurinol and exposed to UV radiation overnight. However, the allopurinol triplet could not be demonstrated in normal eye bank lenses derived from patients who had been on chronic allopurinol therapy for more than 2 years without developing ocular problems.[72] These data suggest that allopurinol can act as a cataractogenic enhancing agent in some patients when it is permanently photobound within their lenses (probably as an additional extrinsically derived photosensitizer). Thus, chronic allopurinol therapy (by itself) does not necessarily result in the retention of allopurinol unless it becomes photobound. The relationship between levels of UV-A exposure and circulating allopurinol levels (and renal function) in the genesis of photosensitized allopurinol cataracts will require further studies.

Experimental cataracts have been reported following tetracycline administration, an effect that may be due to photosensitization.[73] Fraunfelder lists erythemal reactions of the eylids and edema, photosensitivity and erythema multiforme as other ocular side effects of this drug.[74] Its action spectrum ranges between 350 to 420 nm. Since its mechanism of action is poorly understood, it is frequently labeled as phototoxic and possibly photoallergic. Acute transient myopia, blurred vision, diplopia, and papilledema, though rare and for the most part reversible, have also been associated with the administration of tetracyclines. The combined administration of tetracycline and minocycline can result in pseudotumor cerebri.[74] Since extraocular paresis and/or paralysis and·papilledema can be caused by pseudotumor cerebri, these signs and symptoms are probably secondary to the pseudotumor and not directly related to any photosensitizing or toxic effects of the drug.

Other drugs with photosensitizing photoallergic and/or phototoxic properties include the sulfonamides, the oral hypoglycemic agents, antimalarial agents (chloroquine), and some of the oral contraceptives.[74]

The increasing interest in using hematoporphyrin derivatives (HPD) for phototherapy (and photodiagnostic procedures) merits careful evaluation regarding their phototoxic (or sensitizing) poten tial. The porphyrin derivatives absorb over a wide range in the UV and visible region of the electromagnetic spectrum and the fact that they appear to be slowly metabolized (HPD can be retained for 3 months after ingestion[75]) necessitates that adequate precautions be taken when such therapy is employed. These compounds can exert a photosensitizing action via the type I and type II reactions. The photodynamic action is mediated via singlet oxygen and has been shown to polymerize lens proteins in vitro.[76] The porphyrin-induced photochemical reactions also involve OH radicals and have been implicated in the generation of H_2O_2,[75] thus the potential for ocular damage certainly exists. Patients undergoing HPD therapy must be monitored for ocular side effects, and they should also be informed about the potential for severe sunburns, particularly within the first 3 months following therapy.

There has been a resurgence of interest in the therapeutic uses of topical vitamin A acid (retinoic acid) and closely related compounds,[77–79] particularly for skin conditions such as acne and related disorders and psoriasis. In addition, such compounds have been tested as a method for treating corneal xerophthalmia, an ocular condition caused by a severe vitamin A deficiency.[80,81] Considerable success has been claimed with such a therapeutic regime; however, the oral administration of isotretinoin has resulted in some side effects; these include Blepharoconjunctivitis or Meibomianitis (33%) dry eyes (20%), contact lens intolerance (8%) and corneal opacities (5%). In addition, cataracts have now been reported in all 11 patients, but the potential cataractogenic action of the Retinoid group of drugs requires further study.[82] Topical retinoic acid in treating experimental corneal xerophthalmia has met with mixed results. Although little is known about the specific mode of action of such drugs, the possibility of phototoxicity (and/or photosensitization) should be considered, particularly with respect to the eye. Also, such drugs (which are analogues of retinoic acid) could be incorporated into the rod photoreceptor elements during the continuous process of outer disc shedding and renewal. This might explain the complaints of impaired night vision, which can occur in some patients receiving these drugs. This side effect clears up rapidly, provided the drug is immediately discontinued. Patients undergoing clinical trials with these drugs should have careful ocular examinations prior to instituting therapy and be reevaluated at specific intervals to assess their ocular status.

The aldose reductase inhibitors have received a considerable amount of attention because of their potential use in retarding certain types of sugar cataracts.[83] Although they are effective in the laboratory, such drugs can undergo photochemical reactions upon exposure to specific wavelengths of UV radiation,[84] that is, they are capable of undergoing phosphorescence. Such triplets have a sufficiently long lifetime (measured in seconds compared with milliseconds to microseconds for fluorescence) that they can generate photochemical reactions in biologic tissue, thereby acting as photosensitizers. There is some experimental evidence that these drugs can be photobound within the lens,[84] and the possibility that such photobinding could be of significance in the lens and retina in the young eye, or aphakic and pseudophakic eye, requires further investigation and evaluation.

Clinical Studies

Since laboratory studies have demonstrated enhanced fluorescence in the ocular lens associated with aging and PUV-A therapy and PUV-A cataracts have recently been reported, a method to monitor lens fluorescence in vivo has been developed. A new slit lamp densitographic apparatus (based on the Scheimpflug principle) capable of accurately and reproducibly recording visible changes in lens density as it ages was recently introduced.[85,86] We have modified this apparatus to utilize UV radiation (300–400 nm) to measure and quantitate age-related fluorescence levels in normal lens in vivo and to correlate them with our previously reported in vitro data.[87–89] Representative visible and UV slit lamp photographs (taken with the Scheimpflug Topcon SL45 camera) of normal eyes and corresponding densitograms show increased lens fluorescence with age. A series of UV and visible slit lamp photographs of normal patients ranging in age from 5 years to 65 years are shown in Figures 10.3 to 10.5. Note particularly the lack of fluorescence in the young lens and the progressive increase in fluroescence with age. These data can be expressed in graphic form (Fig. 10.6) showing the normal age-related increase in lens fluorescence (in vivo), which corresponds well with the in vitro data (Fig. 10.7 and 10.8) that we previously reported.[1,14]

Aside from demonstrating the normal age-related increase in lens fluorescence, one can also detect abnormally enhanced fluorescence caused by occupational (or accidental) exposure to higher levels of UV radiation. This is shown in Figure 10.9, which is a photograph of a 40-year-old patient who was exposed to excessive UV in his workplace. The increased fluorescence can easily be appreciated by comparing this lens with a photograph of a normal 40-year-old eye (Fig. 10.10). Enhanced fluorescence and/ or abnormal fluorescence emission can also occur in patients on PUV-A therapy, and failure to properly protect such patients from all UV radiation (for at least 24 hours following ingestion of the

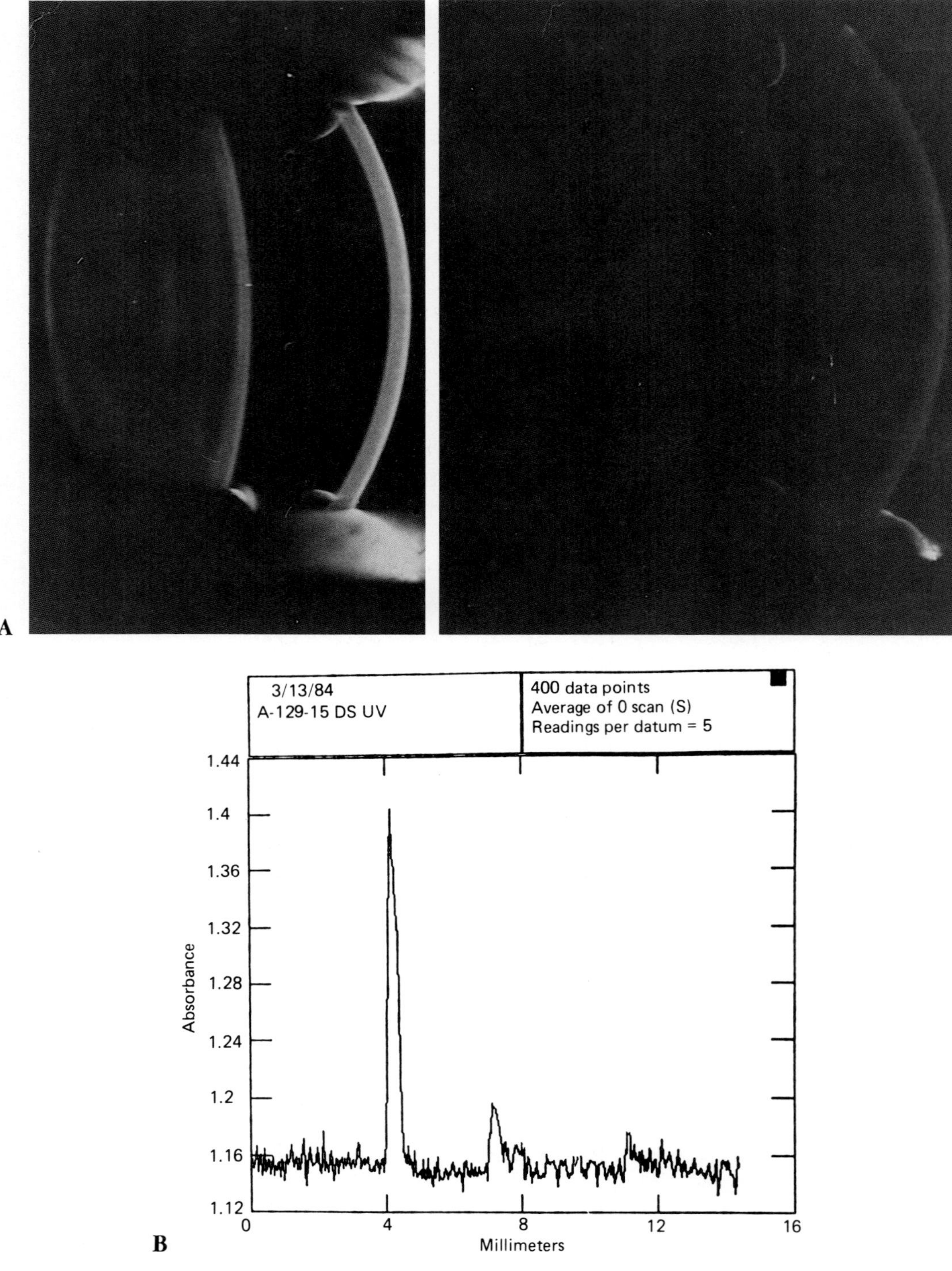

Figure 10.3. **A.** Visible (L) and UV (R) photographs of a 5-year-old normal eye; **B.** UV (fluorescence) densitogram.

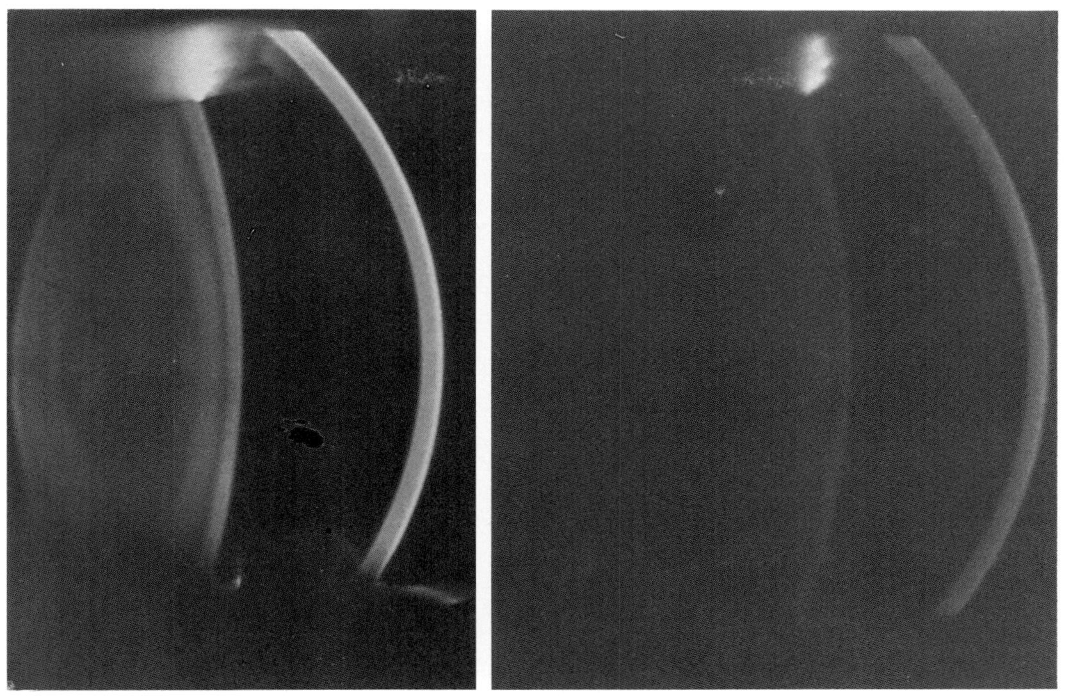

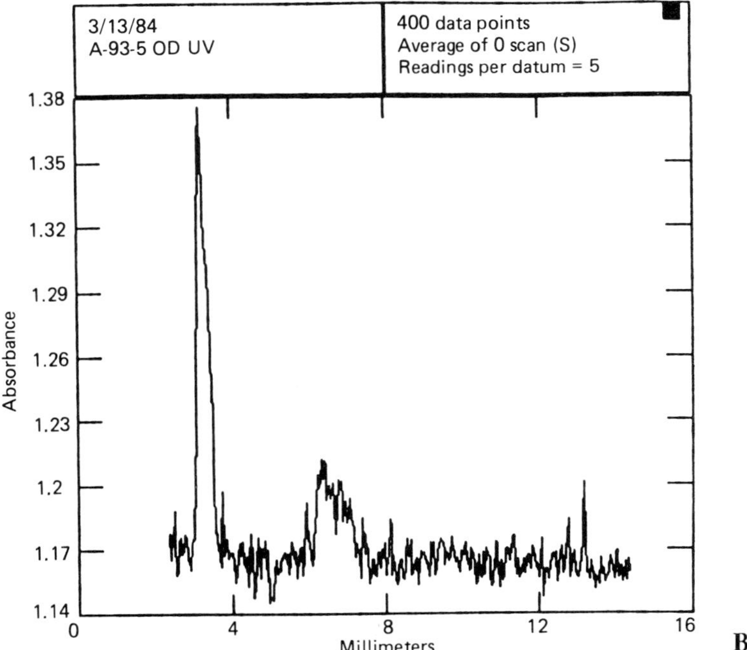

Figure 10.4. **A.** Visible (L) and UV (R) photographs of a 28-year-old normal eye; **B.** UV (fluorescence) densitogram.

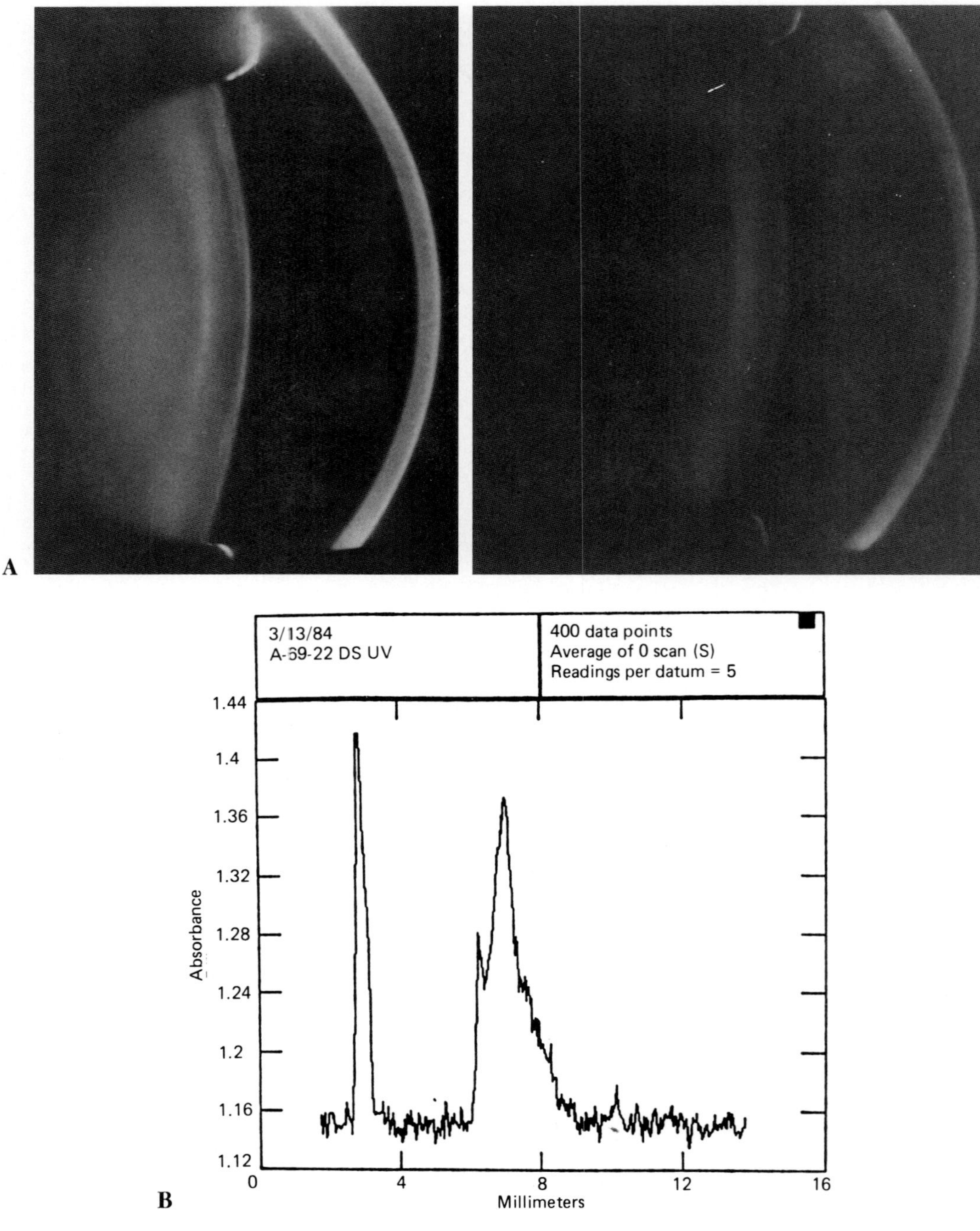

Figure 10.5. **A.** Visible (L) and UV (R) photographs of a 65-year-old normal eye; **B.** UV (fluorescence) densitogram.

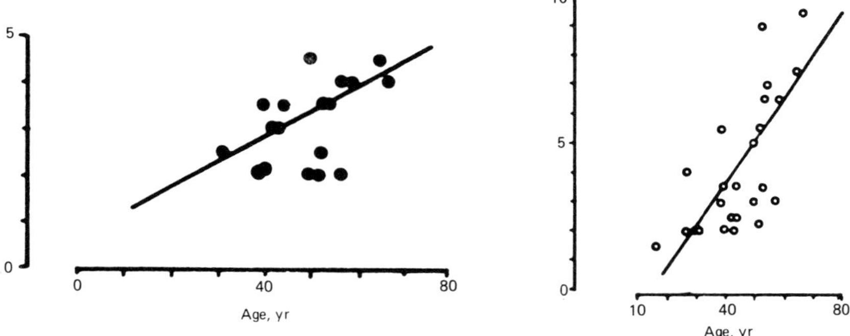

Figure 10.6. Normal age-related increase in two densitographic regions (derived from ultraviolet slit lamp photos in vivo) that correspond to the 440-nm (Fig. 10.7) and 520-nm (Fig. 10.8) fluorescence emission levels obtained in vitro.

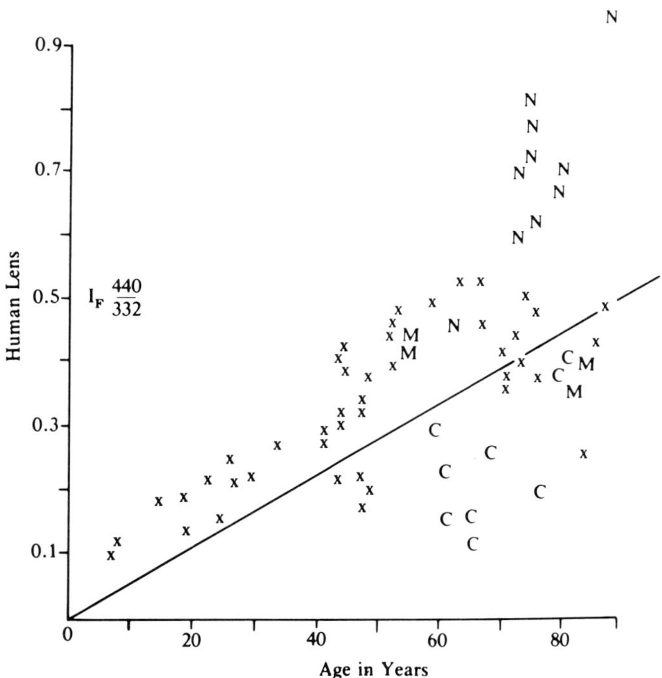

Figure 10.7. I_F440/332 ratios representing whole lens fluorescence intensity at 440 nm (360 nm excitation) divided by tryptophan intensity (in whole lens) at 332 nm (290 nm excitation). The I_F ratio shows age-related increase in the normal lens (X and solid line), a marked increase in brown nuclear cataracts (N), relatively normal or below normal levels in cortical cataracts (C), and high normal values in mixed cortical and nuclear cataracts (M). Each point represents a single lens.

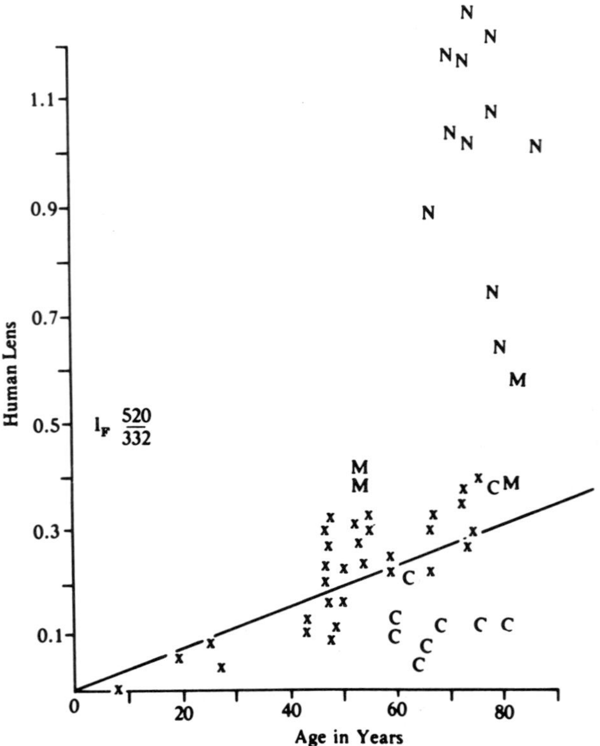

Figure 10.8. I_F 520/332 ratios representing fluorescence intensity of second fluorescent region in the lens at 520 nm (420–435 nm excitation) divided by tryptophan fluorescence intensity in the lens at 332 nm (295 nm excitation). Interference filters (295 and 435 nm) were used to decrease the light scattering when cortical and mixed cataracts were examined. Each point represents a single lens.

drug) can result in cataract formation as shown in Figure 10.11. This 52-year-old patient with psoriasis received 4 years of intermittent PUV-A treatment (without proper eye protection). We have now seen three such PUV-A cataracts and several others have recently been reported in the literature. Although our dermatology clinic now provides all PUV-A patients with proper UV-absorbing or reflecting spectacles, data on a series of such patients who were treated prior to 1977 (when the potential for photosensitized lens damage from psoralen therapy was first demonstrated) and their densitograms demonstrate a significant elevation of one of the lens fluorescence peaks (Fig. 10.12). Patients who have been on D-penicillamine therapy (for a variety of diseases) tend to have lower lens fluorescence intensities (Fig. 10.12). We anticipated this finding since we had previously demonstrated that D-penicillamine (which is an excellent free radical scavenger as well as a chelating agent) is capable of entering the lens, as shown by in

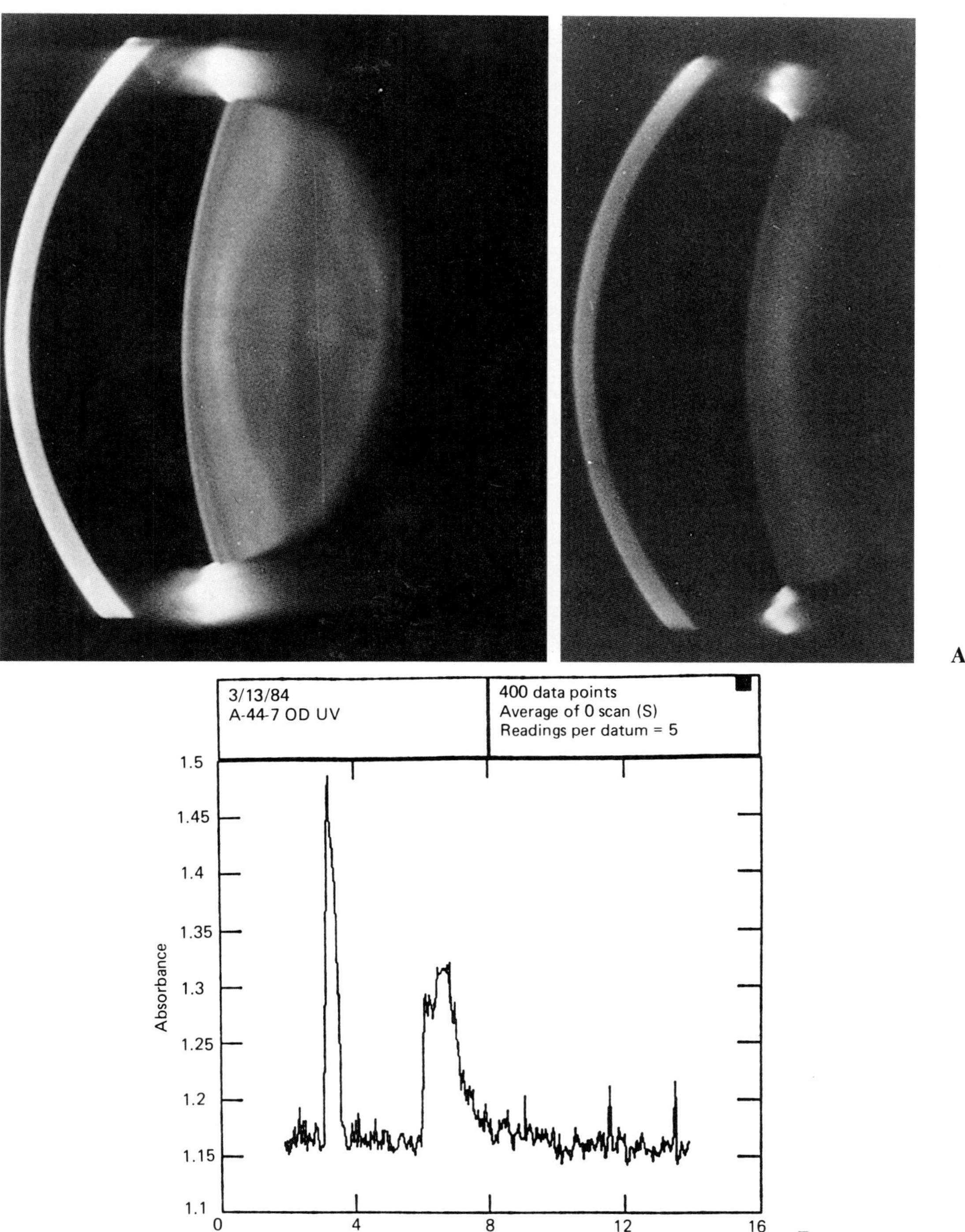

Figure 10.9. **A.** Visible (L) and UV (R) photos of a 40-year-old eye showing abnormal fluorescence due to excessive occupational UV exposure; **B.** UV (fluorescence) densitogram. Note enhanced fluorescence compared with normal values (see Fig. 11.10B).

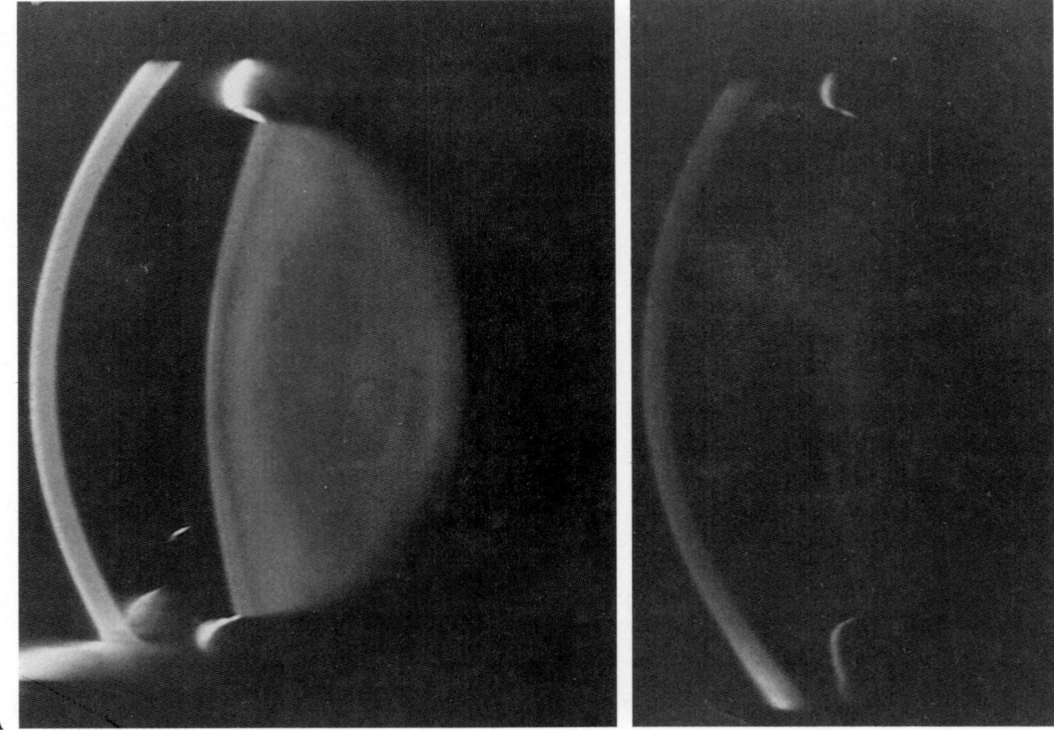

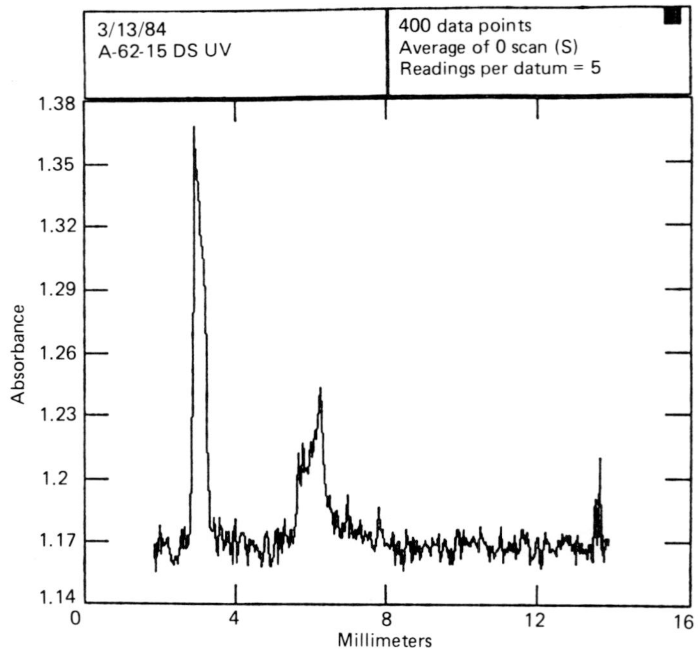

Figure 10.10. **A.** Visible (L) and UV (R) photographs of a normal 40-year-old eye; **B.** UV (fluorescence) densitogram.

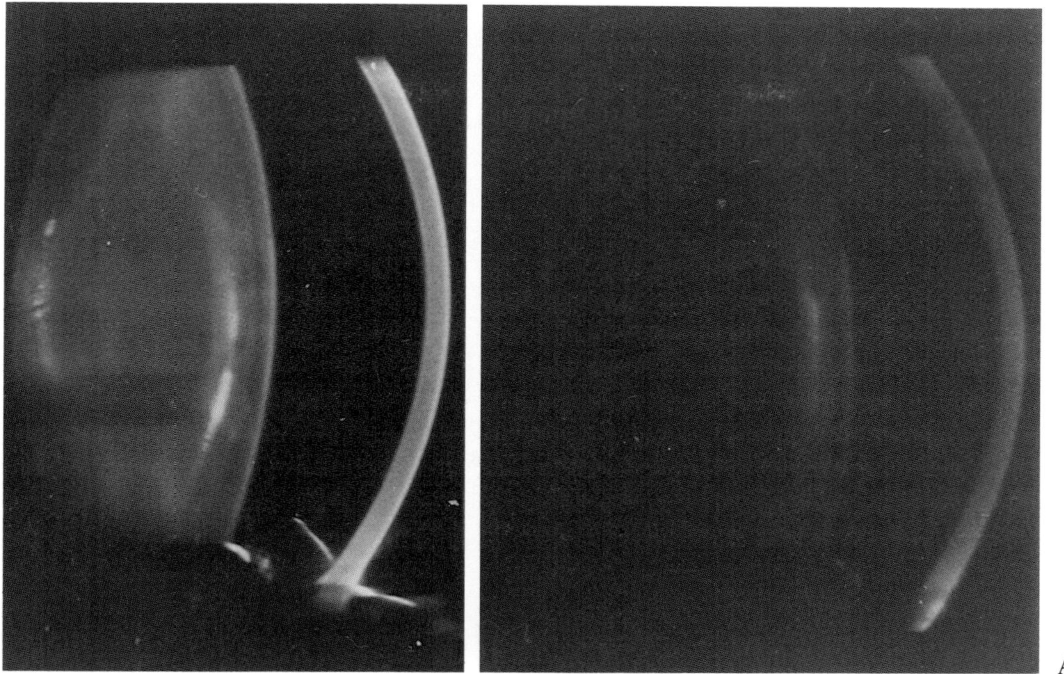

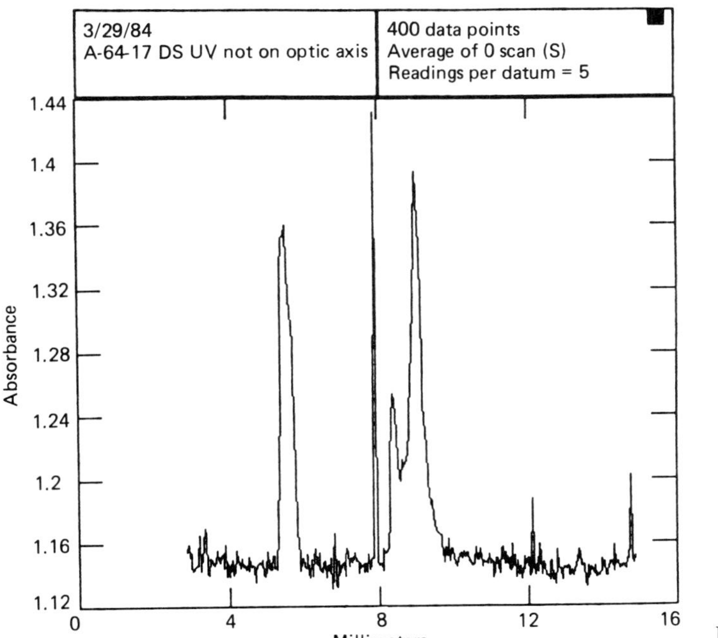

Figure 10.11. **A.** Visible (L) and UV (R) photographs of a PUV-A induced cataract with abnormal fluorescence in a 52-year-old patient on PUV-A therapy for 4 years; **B.** UV (fluorescence) densitogram.

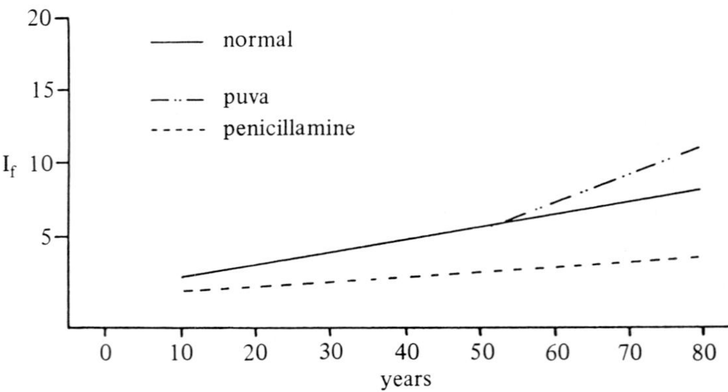

Figure 10.12. Densitographic analyses from in vivo photographs showing enhanced fluorescence in PUV-A patients compared with normal individuals. Note decreased fluorescence level in patients on chronic D-penicillamine (over 1 year) therapy.

vivo as well as in vitro experiments. As a free radical scavenger, D-penicillamine aborts the UV-induced free radical reaction, thereby preventing photodamage.

These studies demonstrate the feasibility of obtaining in vivo lens fluorescence data that are objective, reproducible, and can be quantified. Thus, UV slit lamp densitography can be used to objectively monitor one parameter of lens aging (fluorescence), as well as photosensitized lens damage, at a molecular level years before visible opacities become manifest by conventional slit lamp examination, and measures can be insitituted to retard or prevent UV-induced (direct or photosensitized) lens opacities.

Aside from detecting abnormal levels (or wavelengths) of lens fluorescence in the living eye, we have also tested the hypothesis that the UV-filtering capacity of the lens (which can be directly correlated with increasing fluorescence levels) is lower than normal in certain patients with degenerative retinal disease. That is, the ocular lens in such individuals has not developed sufficient chromophores to enable it to absorb all the UV radiation. In 90% of our population, by the time they reach the third to fourth decade, the lens has become an effective filter for UV and short-wave visible radiation (320–450 nm), thereby protecting the aging (and metabolically less efficient) retina from potential damage. Such wavelengths can cause retinal photodamage as demonstrated by experiments in aphakic primates. We can now determine whether this occurs in the human by measuring lens fluorescence levels. Preliminary results demonstrate a significantly lower level (30%–50%) of lens fluorescence in patients with retinal degenerative diseases (compared with the usual values for their age groups), indicating that their lenses are less effective filters for the 320 to 450 nm wavelengths of radiation. These data suggest that photo-damage may be a significant factor in the progression (and perhaps pathogenesis) of some retinal degenerative diseases and could be particularly relevant for the older pseudophakic patient whose clear PMMA intraocular lens (IOL) is a good transmitter of UV-B as well as UV-A radiation.

In addition to performing in vivo lens fluorescence measurements, the same photographs can also be used (with proper software) to obtain biometric measurements[90,91] on these eyes, including 1) radius of curvature of the anterior and posterior cornea and corneal thickness; 2) depth of the anterior chamber; and 3) radius of curvature of the anterior and posterior lens surfaces and lens thickness. Thus the clinician can now obtain reproducible biometric data on the anterior segment ocular tissues and structures by simply using the Schiempflug slit lamp photographic method. These data will provide an accurate method for monitoring age-related changes in the normal eye (e.g., AC (anterior chamber) diameter, lens thickness, lens density, growth of the lens nucleus, and increased lens fluorescence). The progress of lenticular opacification can also be followed using visible light densitography, and lenticular and corneal fluorescence data can be obtained concurrently by substituting UV radiation for the visible exciting light with appropriate filters.

Ophthalmolscopy, the Operating Microscope, and Retinal Damage

One should also consider the potential problem regarding indirect ophthalmoscopy and retinal damage. Several investigators have demonstrated that indirect ophthalmolscopy is capable of producing retinal damage in primate eyes.[92,93] Indirect ophthalmoscopy involves the use of a 20-diopter convex lens placed between the light source and the patient's eye, which in effect serves as an additional focusing element, thereby further concentrating the light and energy per unit area onto the patient's eye. It has been estimated that the amount of focal energy applied to the retina and choroid by the indirect ophthalmoscope is approximately 0.1 to 0.2 mW/cm^2, which approximates the amount of irradiance received at the retinal surface from the sun. However, the indirect ophthalmoscope has much of its power in the infrared region (longer than 750 nm). Retinal irradiance levels with an indirect versus a direct ophthalmoscope have been measured, and the data indicate a tenfold increase in retinal irradiance with the indirect ophthalmoscope. It is estimated that at the patient's retina the total irradiance is composed of one-third visible light to two-thirds infrared radiation (750–1,400 nm). It has been proposed that infrared filters should be incorporated into all ophthalmoscopes to prevent potential thermal damage from extensive indirect ophthalmoscopy and that reasonably short exposures should be used when examining the posterior pole of the eye. Intense illumination and frequent reexamination at short intervals should be avoided. It should be pointed out that the energy delivered during indirect ophthalmoscopy is only 200 times less than the energy capable of producing a retinal burn.

Recent reports have described clinically demonstrable retinal lesions in patients undergoing intraocular surgery with the operating microscope.[34,35] Lesions resembling central serous retinopathy have been demonstrated in primate eyes exposed to long-wave UV radiation. One must therefore consider the potential of retinal damage with these clinical techniques, particularly in certain conditions such as retinitis pigmentosa, which might be accelerated by light exposure. Photic damage has also been shown to result in pathologic changes in the pigment epithelium as well as in the outer segments; this suggests the possibility that such damage could weaken the adhesions between the retina and pigment epithelium and potentiate retinal detachment.

Summary and Conclusions

The levels of artifical illumination that we are exposed to at present vary between 20 and 50 footcandles (3.2 to 8.1 $\times$ 10^{-5} W/cm^2). Although these levels of irradiance are well below the levels that are capable of producing retinal photopathology in primates and man, they have been shown to be of sufficient intensity to produce permanent as well as reversible retinal damage in certain experimental animals, particularly albino rats. The trend by lighting engineers toward higher levels of illumination (e.g., 100–1,000 foot-candles) should be approached with extreme caution in view of the potential photic damage that could be incurred at these levels of illumination. The lowest reported retinal damage threshold (to visible light) in primates occurs at wavelengths in the blue region of the spectrum, and the potential adverse effects of UV radiation on the retina must also be considered, particularly in the aphakic and pseudophakic eye. It should also be noted that the retina is at least six times more sensitive to UV-A radiation (320–400 nm) than to blue light. Our intraocular lens serves as an efficient filter to remove these wavelengths but when it is extracted, the aphakic (or pseudophakic) retina is exposed to longer wavelength ultraviolet radiation and to blue light. This radiation is capable of exerting significant photochemical damage in man and experimental animals. Such damage can be induced even with chronic cumulative exposure to ambient levels as well as acute, high-intensity irradiance.

One must also consider the role of photosensitizing agents in retinal photobiology. The fact that the psoralens absorb mostly in the longer wavelength UV spectrum (between 320 and 360 nm) has allayed concern about the risk of this drug to the retina. However, the ocular hazard from such photosensitizing drugs should be of concern in aphakic eyes that have lost their natural UV filter (the ocular lens), in pseudophakic, and in young children (whose own lenses have yet to develop as an effective UV filter).

Even if current surgical procedures (operating room microscope, indirect ophthalmoscopy, etc) are not inducing clinically manifest irreversible retinal damage, they are of sufficient intensity when employed for the prolonged time periods frequently required (about 1–3 hours) to cause concern. Since proper precautions can easily prevent this potential photodamage, the surgeon is urged to reduce such hazards by employing proper filters (absorbing all the UV and short wavelength visible radiation up to 475 nm) in his operating room microscope. Similar filters could also be used with indirect ophthalmoscopy. Since much of the optical radiation we employ contains a significant portion of infrared radiation (longer than 750 nm), a second filter to remove these wavelengths is also advocated. As previously noted, although such infrared radiation exposure by itself may be harmless, it could result in sufficient localized heating to the retinal area being illuminated, resulting in the enhancement of photochemical reactions caused by the shorter wavelength radiation (i.e., thermal enhancement of photochemical damage).

Aside from preventing photochemical damage to the retina, the use of UV and short wavelength absorbing filters (300–475 nm) might provide an additional advantage. It is well known that removing wavelengths below 450 nm improves the image quality in ophthalmic photography (by reducing chromatic aberration and light scattering) with only minimum effects on color balance. A similar result would pertain to retinal viewing with indirect ophthalmoscopy when such filters are used.

The current consensus appears to be that aphakic and pseudophakic patients (the vast majority of these patients have IOLs that are excellent transmitters of UV radiation) should be cautioned about potential retinal photodamage. In particular, they should be warned about viewing midday sunlight, welding arcs, and the potential for cumulative retinal photodamage during leisure activities such as boating, snow skiiing, mountain climbing, sunbathing, etc. Since the numbers of vigorous and active people with pseudophakia are rapidly increasing (with more surgeons implanting IOLs in people well below the fifth or sixth decade of life), such individuals have a potential life span sufficient to accumulate retinal photodamage, which generally requires two or more decades before becoming manifest. Until a decade or two ago, there was a tendency to operate only on patients with advanced cataracts. The patients in the older groups would wait until their cataracts reached a point we called "ripe." Therefore, we had patients whose postoperative life span was much shorter, whereas with the advent of IOL surgery, we now have patients with a much more prolonged life span. In previous years, we may have prevented such patients access to chronic cumulative aphakic UV exposure by operating on their cataracts at a point where the patient's longevity was limited. As intraocular lenses are inserted into patients who are much younger and will live longer, we must be aware of the possibility that retinal degenerative diseases could increase during the next decade unless all aphakia and pseudophakia patients are properly protected from chronic (long term) exposure to UV-A and short wavelength visible radiation (300–450 nm) by appropriate filters.

Because of the confusion and controversial claims regarding the efficacy of commercially available sunglasses in protecting the eye from UV photodamage, we have analyzed a large series of such lenses to determine their transmission characteristics.[94] Their ability to transmit UV and visible radiation (280–750 nm) was measured in a recording spectrophometer. Photon flux measurements were also performed, with the lens exposed to a 200 W Hg light source filtered to transmit only 280 to 390 nm radiation. These studies are in general agreement with an earlier report [95] and demonstrate a wide variation in the UV transmission characteristics of sunglasses evaluated, ranging from 1.5% to 40%, with similar transmission values noted when tested for more discrete wavelengths (340–380 nm). Only the NOIR, Spectra-Shield, Silor, Univis, and UV 400 lenses were >99% effective in filtering the UV radiation. It should also be noted that visible radiation is significantly decreased in darkly tinted sunglasses, while still permitting some long wavelength UV transmission. In patients with blue, gray, and hazel eyes, a 50% or more decrease in visible radiation can result in pupillarly enlargement of up to 0.25 mm, thereby increasing the effective dose of radiation incident on the intraocular tissues.

Therapy

The best and simplest treatment for direct UV photodamage to ocular tissues is prevention. Spectacle lenses that are excellent UV filters have recently been introduced; these include a variety of UV-absorbing plastic lenses. Of the latter, the Silor, Univis, and UV 400 are equally efficient filters and can be ordered with the patient's correction. For those who do not require corrective lenses, plano spectacles made of any of the foregoing materials can be ordered. For patients on phototherapy, goggles that include side pieces are preferred (to remove reflected radiation). Ordinary commercial sunglasses are not necessarily effective absorbers of UV radiation longer than 320 nm and are not recommended unless their transmission characteristics are such that they remove 99+% or all UV radiation.

Despite recent claims by some IOL manufacturers that their lenses absorb UV radiation, there are as yet no proven products available. Since clear glass absorbs UV up to 320 nm, and some intraocular lenses are made of materials that contain some absorbing chromophores, manufacturers can claim that their lenses are UV filters even though they do not remove the longer wavelengths. To be truly effective, the IOL must filter all radiation up to 400 nm and a significant percentage of the shorter wavelength visible light (400–450 nm). Such lenses will probably become available in the near future. It is hoped that any lenses claiming to be UV filtering IOLs will have their true absorption and transmission characteristics clearly noted as a package insert. It should also be noted that all the PMMA IOLs currently on the market leach small (microgram) quantities of free methacrylate. Fortunately, these low levels do not cause any problem, and the lenses appear to be well tolerated. However, the recent introduction of the Nd:YAG laser to perform posterior capsulotomies in patients who develop secondary posterior lens capsule opacification may give rise to other problems. Although the Nd:YAG laser certainly performs capsulotomies effectively, there is now evidence that a significant number of IOLs are "pitted" during this procedure. Studies in our laboratory have demonstrated a 20-fold or greater increase in the amount of methacrylate leached from such laser-pitted lenses.[96] This marked increase could cause a recurrence of intraocular reactions—such as chronic intraocular inflammatory reactions—similar to those experienced when PMMA IOLs were first introduced. This problem was attributed to methacrylate leaching and was solved by improvements in the manufacture of the template material. Furthermore, all the UV-absorbing IOLs currently available leach small but measurable amounts of chromophore, and the amount leached is also markedly increased when these lenses are pitted with Nd:YAG laser. The long-term consequences of increased leaching of organic chromophores (some of which are potential photosensitizers) could be significant, resulting in, for example, the development of a low-grade cyclitis or uveitis and even the permanent incorporation of photosensitizers within the neural retina. Further studies and careful long-term clinical follow-ups should be instituted to evaluate this problem.

References

1. Lerman S: Radiant Energy and the Eye, Chapters 1–3, MacMillan Publishing Company, New York, 1980.
2. Kurzel RB, Wolbarsht ML, Yamanashi BS: UV radiation effects on the human eye. Photochem Photobiol Rev 2:133–167, 1977.
3. Lerman S, Tan TT, Louis D, Hollander M: Anomalous absorption of lens proteins due to a fluorogen. Ophthalmic Res 1:338–343, 1970.
4. Zigman S: Eye lens color formation and function. Science 171:807–809, 1971.
5. Lerman S: Lens proteins and fluorescence, Isr J Med Sci 8:1583–1589, 1972.
6. Pirie A: Effect of sunlight on proteins of the lens, in Bellows J (ed): Contemporary Ophthalmology, Williams and Wilkins, Baltimore, pp 484–501, 1972.
7. Satoh K, Bando M, and Nakajima A: Fluorescence in human lens. Exp Eye Res 16:167–172, 1973.
8. Augusteyn RC: Human lens albuminoid. Jap J Ophthalmol 18:127–134, 1974.
9. Dilley KJ, Pirie A: Changes to the proteins of the human lens nucleus in cataract. Exp Eye Res 19:59–72, 1974.
10. Augusteyn R: Distribution of fluorescence in the human cataractous lens. Ophthalmic Res 7:217–224, 1975.
11. Spector A, Roy O, Stauffer J: Isolation and characterization of an age-dependent polypeptide from human lens with non-tryptophan fluorescence. Exp Eye Res 21:9–24, 1975.
12. Bando M, Nakajima A, Satoh K: Coloration of human lens protein, Exp Eye Res 20:489–492, 1975.
13. Lerman S: Lens fluorescence in aging and cataract formation. Doc Ophthalmol Proc Series 8:241–260, 1976.
14. Lerman S, Borkman RF: Spectroscopic evaluation and classification of the normal aging and cataractous lens. Ophthalmic Res 8:335–353, 1976.
15. Lerman S, Borkman RF: Photochemistry and lens aging, in von Hahn HP (ed): Interdisciplinary Topics in Gerontology: Gerontological Aspects of Eye Research, vol 13, S Karger, Basel, pp 154–183, 1978.
16. Lerman S, Kuck JF, Borkman R, Saker E: Induction, acceleration and prevention (in vitro) of an aging parameter in the ocular lens. Ophthalmic Res 8:213–226, 1976.

17. Zigman S, Datiler M, Torozynshi E: Sunlight and human cataract. Invest Ophthalmol Vis Sci 18:462–467, 1979.

18. Castineiras SG, Dillon J, Spector A: Effects of reduction on absorption and fluorescence of human lens proteins. Exp Eye Res 29:573–575, 1979.

19. Yu NT, Kuck JFR, Askren CC: Red fluorescence in older and brunescent human lenses. Invest Ophthalmol Vis Sci 18:1278–1280, 1979.

20. Lerman S: Lens transparency and aging, in Regnault F, Hockwin O, Courtois Y (eds). Aging of the Lens. Elsevier/North Holland Biomedical Press, New York/London pp 263–279, 1980.

21. Garner MH, Spector A: Selective oxidation of cysteine and methionine in normal and senile cataractous lenses. Proc Natl Acad Sci USA 77:1274, 1980.

22. Borkman RF, Dalrymple A, Lerman S: Ultraviolet action spectrum for fluorogen production in the ocular lens. Photochem Photobiol 26:129–132, 1977.

23. Borkman RF: Ultraviolet action spectrum for tryptophan destruction in aqueous solution. Photochem Photobiol 26:163–166, 1977.

24. Borkman RF, Lerman S: Evidence for a free radical mechanism in aging and UV Irradiated ocular lenses. Exp Eye Res 25:303–309, 1977.

25. Zigman S, Vaughn T: Near UV light effects on the lenses and retinas of mice. Invest Ophthalmol 13:462–465, 1974.

26. Pitts DG, Hacker PD, Parr WH: Ocular Ultraviolet Effects from 295 nm to 400 nm in the Rabbit Eye. DHEW (NIOSH) Publication No. 77–175, October, 1977.

27. Lerman S: Human UV radiation cataracts. Ophthalmic Res 12:303–314, 1980.

28. Hiller R, Giacometti L, Yuen K: Sunlight and cataract; An epidemiologic investigation, Am J Epidemiol 105:450, 1977.

29. Taylor HR: The environment and the lens, Br J Ophthal 64, 303, 1980.

30. Hollows F, and Moran D: Cataract—the ultraviolet risk factor. Lancet 1249, Dec 5, 1981.

31. Brilliant LB, Grasset NC, Pokhrel RP, Kolstad A, Lepkowski JM, Brilliant GE, Hawks WN: Associations among cataract prevalence, sunlight hours and altitude in the Himalayas. Am J Epidemiol 113: 250, 1983.

32. Ts'o MOM, Fine BS, Zimmerman LE: Photic maculopathy produced by indirect ophthalmoscope. I. Clinical and histopathic study, Am J Ophthalmol 73, 686, 1972.

33. Ham WT, Muller HA, Ruffolo JJ, Guerry D, Guerry RK: Action spectrum for retinal injury from near ultraviolet radiation in the aphakic monkey, Am J Ophthalmol 93, 299, 1982.

34. Hochheimer B: A possible cause of chronic cystic maculopathy: The operating microscope, Ann Ophthalmol 13, 153, 1981.

35. Berler D, Peyser R: Light intensity and visual acuity following cataract surgery. Ophthalmology 89 [suppl]:117, 1982.

36. Lerman S, Gardner K, Megaw J, Borkman R: Prevention of direct and photosensitized UV radiation damage to the ocular lens. Ophthalmic Res 13:284–292, 1981.

37. Thomas DM, Schepler KL: Raman spectra of normal and ultraviolet-induced cataractous rabbit lens. Invest Ophthalmol Vis Sci 19:904–912, 1980.

38. Noell WK, Albrecht R: Irreversible effects of visible light on the retina: Role of Vitamin A. Science 171, 76, 1971.

39. Ts'o MOM: Photic maculopathy in Rhesus monkey; A light and electron microscopic study, Invest Ophthalmol Vis Sci 12:17, 1973.

40. Lanum J: The damaging effects of light on the retina. Empirical findings, theoretical and practical applications, Surv Ophthalmol 22, 221, 1978.

41. Marshall J, Grindle CFJ, Ansell PL, Borwein B: Convolution in human rods: an aging process, Br J Ophthalmol 63, 18, 1979.

42. Clark B, Johnson ML, Dreher R: The effect of sunlight on dark adaptation, Am J Ophthalmol 29:828, 1946.

43. Hecht S, Hendley CD, Ross H, Richmond PN: The effect of exposure to sunlight on night vision, Am J Ophthalmol 31:1573, 1948.

44. Penner R, McNair JN: Eclipse blindness, Am J Ophthalmol 61:1452, 1966.

45. Potts A, Gonasum LM: Toxicology of the eye, in Toxicology, Casarett JJ, Doull J (eds). MacMillan Publishing Co, New York, pp 275–312, 1975.

46. Cloud TM, Hakim R, Griffin AC: Photosensitization of the eye with methoxalen. I. Acute effect. Arch Ophthalmol 64:346–351, 1960.

47. Cloud TM, Hakim R, Griffin AC: Photosensitization of the eye with methoxsalen. II. Chronic effects. Arch Ophthalmol 66:689–694, 1961.

48. Freeman RG, Troll D: Photosensitization of the eye by 8-methoxypsoralen. J Invest Dermatol 53:449–455, 1969.

49. Lerman S, Borkman R: A method for detecting 8-methoxypsoralen in the ocular lens. Science 197:1287–1288, 1977.

50. Lerman S, Jocoy M, Borkman R: Photosensitization of the lens by 8-methoxypsoralen. Invest Ophthalmol Vis Sci 16:1065–1068, 1977.

51. Jose JG, Yielding KL: Unscheduled DNA synthesis in lens epithelium following ultraviolet irradiation. Exp Eye Res 24:113–119, 1977.

52. Jose JJ, Yielding KL: Photosensitive cataractogens, chlorpromazine and methoxypsoralen cause DNA repair systhesis in lens epithelial cells. Invest Ophthalmol Vis Sci 17:687–690, 1978.

53. Lerman S, Megaw J, Willis I: Potential ocular complications of PUV-A therapy and their prevention. J Invest Dermatol 74:197–199, 1980.

54. Lerman S, Megaw J, Willis I: The photoreaction of 8-MOP with tryptophan and lens proteins. Photochem Photobiol 31:235–243, 1980.

55. Megaw J, Lee J, Lerman S: NMR analyses of tryptophan-8-methoxypsoralen photoreaction products. Photochem Photobiol 32:265–270, 1980.

56. Crylin MN, Pedvis-Leftick A, Sugar J: Cataract formation in association with ultraviolet photosensitivity. Ann Ophthalmol 12:786–790, 1980.

57. Lerman S, Megaw J, Gardner K, Takei Y, Willis I: Localization of 8-methoxypsoralen in ocular tissues. Ophthalmic Res 13:106–116, 1981.

58. Wulf HC, Andreasen MP: Distribution of ^{3}H-8-mop and its metabolites in rat organs after a single oral administration. J Invest Dermatol 76:252–257, 1981.

59: Wulf HC, Andreasen MP: Concentration of ^{3}H-8-methoxypsoralen and its metabolites in the rat lens and eye after a single oral administration. Invest Ophthalmol Vis Sci 22:32–36, 1982.

60. Lerman S: Ocular phototoxicity and PUV-A therapy: An experimental and clinical evaluation. FDA Photochemical Toxicy Symp J Natl Cancer Inst 69:287–302, 1982.

61. Lerman S, Megaw J, Gardner K: PUV-A therapy and human cataractogenesis. Invest Ophthalmol Vis Sci 23:801–804, 1982.

62. Lerman S, Megaw J, Gardner K, Takei Y, Franks Y, Gammon A: Photobinding of ^{3}H-8-methoxypsoralen to monkey intraocular tissues. Invest Ophthalmol Vis Sci 25:1267–1274, 1984.

63. Parrish JA, Fitzpatrick TB, Tanenbaum L, pathak MA: Photochemotherapy of psoriasis with oral methoxalen and longwave ultraviolet light. N Engl J Med 291:1207–1211, 1974.

64. Parrish JA, Fitzpatrick TB, Shea C, Pathak MA: Photochemotherapy of vitiligo. Use of orally administered psoralens and a high intensity longwave ultraviolet light (UV-A)system. Arch Dermatol 112:1531–1534, 1976.

65. Lerman S: Psoralens and ocular effects in animals and man: In vivo monitoring of human ocular and cutaneous manifestations. J Natl Cancer Inst Monograph No 66, Photochemotherapeutic Aspects of Psoralens, pp 227–233, 1984.

66. Dayhaw-Barker P, Barker FM II: Retinal effects of short term exposure to 8-MOP and UV-A. Photochem Photobiol 37[Suppl]:S83, 1983.

67. Dayhaw-Barker P, Barker FM II, Diebert K: Effects of three drugs on the retinal threshold to damage. Am J Optom Physiol Opt 5910, 16P, 1982.

68. Meier-Ruge W: Drug induced retinopathy. CRC Crit Rev Toxicol 352, 1972.

69. Baer RL, Harber LC: Photosensitivity induced by drugs. JAMA 192:989, 1965.

70. Fraunfelder FT, Hanna C, Dreis MW, Cosgrove KW: Possible lens changes associated with allopurinol therapy. Am J Ophthalmol 94:137–140, 1982.

71. Lerman S, Megaw J, Gardner K: Allopurinol therapy and human cataractogenesis. Am J Ophthalmol 94:141–146, 1982.

72. Lerman S, Megaw J, Fraunfelder F: Further studies on allopurinol and human cataractogenesis. Am J Ophthalmol 97:205–209, 1984.

73. Krejci L, Brettschneider I, Triska J: Tetracycline hydrochloride and lens changes. Ophthalmic Res 10:30, 1978.

74. Fraunfelder FT: Drug-induced Ocular Side Effects and Drug Interactions. Lea and Febiger, Philadelphia, 1982.

75. Dayhaw-Barker P, Forbes D, Fox D, Lerman S, Metgaw S, McGinniss J, Waxler M, Felten R: Drug Photoxicity and Visual Health. FDA Symposium, Long Term Visual Health Risks of Optical Radiation. Bethesda, Maryland, September 24–27, 1983.

76. Roberts JE: The photodynamic effect of chlorpromazine, promazine and hematoporphrin on lens protein, Invest Ophthalmol Vis Sci 25, 748, 1984.

77. Thomas JR III, Doyle JA: The therapeutic uses of topical vitamin A acid. J Am Acad Dermatol 4:505, 1981.

78. Ward A, Brogden RN, Heel RC, Speight TM, Avery GS: Etretinate, a Review of its pharmacological properties and therapeutic efficacy in psoriasis and other skin disorders. Drugs 26:9, 1983.

79. Shalita AR, Cunningham WJ, Leyden JJ, Pochi PE, Strauss JS: Isotretinoin treatment of acne and related disorders: an uptdate, J Am Acad Dermatol 9:629, 1983.

80. Sommer A, Treatment of corneal xerophthalmia with topical retinoic acid. Am J Ophthalmol 95:349, 1983.

81. Hatchell DL, Faculjak M, Kubicek D: Treatment of xerophthalmia with retinol, tretinoin, and etretinate. Arch Ophthalmol 102, 926, 1984.

82. Fraunfelder FT, LaBraico JM, Meyer SM: Adverse Ocular Reactions Possibly Associated with Isotretinoin. Amer J Ophthalmol 100:534–537, 1985.

83. Lerman S: Observations on the prevention and medical treatment of cataracts, Chapter 51, in Cataract and Intraocular Lens Surgery, Ginsburg SP (ed). Aesculapius Pub Co, Birmingham, Ala, vol 2, pp 671–688, 1984.

84. Lerman S, Megaw J, Gardner K: Optical spectroscopy as a method to monitor aldose reductase inhibitors in the lens. Invest Ophthalmol Vis Sci 24:1505–1510, 1984.

85. Dragomirescu V, Hockwin O, Koch HR: Development of a new equipment for rotating slit image photography according to Scheimpflug's principle, in Interdisciplinary Topics in Gerontology, vol 13, Basel Karger, pp 118–130, 1978.

86. Dragomirescu V, Hockwin O, Koch HR: Photo-cell device for slit-beam adjustment to the optical axis of the eye in scheimpflug photography. Ophthalmic Res 12:78–86, 1980.

87. Lerman S, Hockwin O: UV-visible slit lamp densitography of the human eye. Exp Eye Res 33:587–596, 1981.

88. Lerman S, Dragomirescu V, Hockwin O: In vivo monitoring of direct and photosensitized UV radiation damage to the lens. Acta XXIC Inter Cong Ophthalmol 1:354–358, 1983.

89. Lerman S: Biophysical aspects of corneal and lenticular transparency. Curr Eye Res 3(1):3–14, 1984.

90. Lerman S, Hockwin O: Measurement of anterior chamber diameter and biometry of anterior segment by scheimpflug slit lamp photography. Am Intra Ocular Implant Soc J 11:149–152, 1985.

91. Lerman S, Hockwin O: Automated biometry and densitography of anterior segment of the eye. Graefe Arch Clin Exp Ophthalmol 223:121–129, 1985.

92. Dawson WW, Herron WL: Retinal illumination during indirect Ophthalmoscopy: Subsequent dark adaptation. Invest Ophthalmol 9:89, 1970.

93. Ts'o MOM, Fine BS, Zimmerman LE: Photic maculopathy produced by the indirect ophthalmoscope, I. Clinical and histopathic study. Am J Ophthalmol 73:686, 1972.

94. Lerman S, Megaw J: Transmission characteristics of commercially available sunglasses. J Ocular Cut Toxicol 2:47–61, 1983.

95. Anderson WJ, Gebel RKH: Ultraviolet windows in commercial sunglasses. Appl Opt 16:515–517, 1977.

96. Lerman S: Observations on UV absorbing IOL's. Curr Can Ophthalmic Pract 3:43–47, 1985.

Index